CAUSATION IN DECISION, BELIEF CHANGE, AND STATISTICS

THE UNIVERSITY OF WESTERN ONTARIO SERIES IN PHILOSOPHY OF SCIENCE

A SERIES OF BOOKS IN PHILOSOPHY OF SCIENCE, METHODOLOGY, EPISTEMOLOGY, LOGIC, HISTORY OF SCIENCE, AND RELATED FIELDS

VOLUME 42

CAUSATION IN DECISION, BELIEF CHANGE, AND STATISTICS

Proceedings of the Irvine Conference on Probability and Causation

VOLUME II

Edited by

WILLIAM L. HARPER

Talbot College, The University of Western Ontario

and

BRIAN SKYRMS

Department of Philosophy, University of California at Irvine

KLUWER ACADEMIC PUBLISHERS

DORDRECHT / BOSTON / LONDON

Library of Congress Cataloging-in-Publication Data

Irvine Conference on Probability and Causation (1985: University of California at Irvine)
Causation in decision, belief change, and statistics / edited by William L. Harper and Brian Skyrms.
p. cm. – (Proceedings of the Irvine Conference on Probability and Causation; v. 2) (The University of Western Ontario series in philosophy of science; v. 42)
Includes bibliographies and index.
ISBN 90–277–2634–5
1. Probabilities–Congresses. 2. Causation–Congresses. I. Harper, William L. (William Leonard), 1943– . II. Skyrms, Brian. III. Title. IV. Series. V. Series: Irvine Conference on Probability and Causation (1985: University of California at Irvine). Proceedings of the Irvine Conference on Probability and Causation; v. 2.
BC141.I78 1985 vol. 2
122 s–dc 19
[121'.6]

Published by Kluwer Academic Publishers,
P.O. Box 17, 3300 AA Dordrecht, The Netherlands.

Kluwer Academic Publishers incorporates the publishing programmes of
D. Reidel, Martinus Nijhoff, Dr W. Junk and MTP Press

Sold and distributed in the U.S.A. and Canada
by Kluwer Academic Publishers,
101 Philip Drive, Norwell, MA 02061, U.S.A.

In all other countries, sold and distributed
by Kluwer Academic Publishers Group,
P.O. Box 322, 3300 AH Dordrecht, The Netherlands.

Printed in The Netherlands

TABLE OF CONTENTS

CONTENTS OF VOLUME I

PREFACE

The papers collected here are, with three exceptions, those presented at a conference on probability and causation held at the University of California at Irvine on July 15—19, 1985. The exceptions are that David Freedman and Abner Shimony were not able to contribute the papers that they presented to this volume, and that Clark Glymour who was not able to attend the conference did contribute a paper. We would like to thank the National Science Foundation and the School of Humanities of the University of California at Irvine for generous support.

WILLIAM HARPER
University of Western Ontario

BRIAN SKYRMS
University of California at Irvine

INTRODUCTION

PART I: DECISIONS AND GAMES

Causal notions have recently come to figure prominently in discussions about rational decision making. Indeed, a relatively influential new approach to theorizing about rational choice has come to be called "causal decision theory".[1]

Decision problems such as Newcombe's Problem and some versions of the Prisoner's Dilemma where an act counts as evidence for a desired state even though the agent knows his choice of that act cannot causally influence whether or not the state obtains have motivated causal decision theorists. Such theorists agree that examples of this sort show that epistemic decision theories such as Jeffrey's (1965) V-maximizing approach,

$$V(A) = \sum P(B_i \mid A) \cdot u(A \,\&\, B_i),$$

which use the agent's epistemic conditional probability

$$P(B_i \mid A)$$

for states B_i on act A as the appropriate probability weight for computing expected utility of act A with respect to a partition of states B_i, are not appropriate as general decision rules. Causal theories are characterized by the introduction of additional apparatus designed to allow for more sensitive representation of an agent's beliefs about what he or she can causally influence than can be captured by appeal only to his or her epistemic conditional probabilities.

Robert Stalnaker's (1973) original proposal, and its development by Gibbard and Harper (1978) introduced subjunctive conditionals. On this approach the appropriate expectation is

$$U(A) = \sum_i P(A \mathbin{\Box\!\!\rightarrow} B_i) \cdot u(A, B_i).^2$$

The unconditional epistemic probability of the subjunctive conditional

William L. Harper and Brian Skyrms (eds.), Causation in Decision, Belief Change, and Statistics, II, ix–xix.

with the proposition that act A is done as antecedent and that state B_i obtains as consequent is proposed as the appropriate probability weight for computing expected utility of act A with respect to a partition of states B_i. Howard Sobel (1978, 1986) introduced objective conditional chances and David Lewis (1981) introduced subjunctive conditionals with propositions about objective chances as consequents. One way to formulate a version of such an account uses the epistemic expectation of the objective conditional probability of the states B_i on the act as the appropriate probability weight,

$$U(A) = \sum_i \sum_j P(H_j) \cdot C_j(B_i \text{ given } A) \cdot u(A, B_i).$$

Here the H_j's are alternative hypotheses about the relevant chance set up and

$$C_j(B_i \text{ given } A)$$

is construed as the single case conditional chance or propensity of B_i on A that corresponds to hypothesis H_j. Another approach introduces explicit causal hypotheses about what the agent can (David Lewis, 1981) or cannot (Brian Skyrms, 1980, 1984) causally influence. Skyrms' version of this approach would have

$$U(A) = \sum_j P(K_j) \cdot \sum_i P(B_i \mid K_j \mathbin{\&} A) \cdot u(A, K_j \mathbin{\&} B_i)$$

where the K_j's are an appropriate partition of the causally relevant factors that are causally independent of the agent's choice. Here the expected utility of A with respect to the B_i's is computed conditional on each K_j and the utility of A is the expectation over the K_j's of this evaluation.

It can be argued (e.g. Lewis, 1981) that these various forms of causal decision theory are equivalent — that an adequate version of any one of these approaches will be interdefinable with adequate versions of the others. Think of the K_j's as acting for the agent as alternative specifications of the relevant chance set up so that conditional belief in B_i on A and K_j acts as the agent's assignment of what he or she takes to be the objective conditional chance of B_i on A corresponding to chance hypothesis K_j. Such a construal makes the K-expectation equivalent to Sobel's formulation where the probability weight is an epistemic expectation of the objective conditional chance. If the appropriate epistemic evaluation of a subjunctive conditional

$$A \mathrel{\Box\!\!\rightarrow} B_i$$

is the epistemic expectation of the corresponding objective chance then the Stalnaker formulation developed by Gibbard and Harper is equivalent as well.[3]

Subjunctive conditionals and objective single case propensities or chances involve problems, obscurities and pragmatic dimensions that have made them difficult to explicate just as has been the case with the idea of causation or causal influence itself. One main attraction of epistemic decision theory has been its promise to avoid introducing these problematic notions into an account of rational decision making. Epistemic conditional probability has seemed to be far less problematic than these notions that causal decision theorists have introduced into the discussion. The main case for causal decision theory has been argued by use of specific versions of Newcombe's Problem (e.g. the Jones example, Harper, 1980, pp. 16—18) where causal independence seems clearly to hold (and to be relevant) even if we have no general account of causal notions that can rival our accounts of epistemic conditional probability. Such cases have seemed to causal decision theorists to be clear cut enough to show the inadequacy of epistemic decision theory and to make accounts which introduce more problematic causal notions worth exploring.

Representation theorems for classic subjective probability (e.g. Savage, 1954; Herstein and Milnor, 1953; Fishburn, 1973; and a host of others) and for epistemic decision theory (e.g. Bolker, 1967) have provided a foundation for the idea of epistemic conditional probability based on relatively intuitive qualitative constraints or rational preference. These foundational results have also offered hope for methods whereby one could, at least in principle, measure epistemic conditional probabilities by appealing to relatively accessible comparative preferences. Some of the idealizations (e.g. Levi, 1980, 1984, 1986) and even some of the most fundamental preference axioms (e.g. Machina, 1982) involved in these foundational results have come under attack. Moreover, empirical work (e.g. Kahneman *et al.*, 1982) has shown that the measurement of degrees of conditional belief (or even unconditional belief) is far less straight-forward than many had hoped. Nevertheless, the representation theorems for epistemic decision theory have been rightly cited (e.g. Eells, 1982, p. 148) as a significant advantage over causal theories which had no comparable foundational results.

One specific virtue simplifying the foundational problem for epistemic decision theory is that its decision rule is formulated for any arbitrary partition of states B_i. The additional difficulties introduced by the problematic notions causal decision theorists appeal to are nicely illustrated by the problem of finding an adequate partition of states K_j for Skyrms' formulation.

In the first paper of this volume, "Conditional Preference and Causal Expected Utility", Brad Armendt appeals to illuminating ideas about qualitative conditional preference to give a lucid account of which partitions of states and consequences are appropriate for a Skyrmsian causal decision theoretic formulation of any given decision problem. He then shows how to embed this account in a version of Peter Fishburn's axiomatic framework for conditional preference to produce the first fully worked out representation theorem for causal decision theory.

A common objection to causal decision theory is that the cases used to motivate its problematic notions are so unusual that they can be ignored without sacrificing any important applications of decision theory. The second paper in this volume, William Harper's "Causal Decision Theory and Game Theory: A Classic Argument for Equilibrium Solutions, a Defence of Weak Equilibria, and a New Problem for the Normal Form Representation", argues that causal decision theory can illuminate a number of important (and widely disputed) foundational issues in game theory. If this is correct then the motivation for introducing problematic causal notions into decision theory is far more extensive than a few strange examples like Newcombe's Problem.

The third paper, Ernest Adams' "Consistency of Decision: Variations on Ramseyan Themes'', re-investigates some of the deep questions raised by Frank Ramsey's (1926) pioneering classic "Truth and Probability". Adams suggests that many contemporary researchers on rationality have overlooked important aspects of these questions. He offers interesting investigations of the idea that appropriate frequencies can play the role for degrees of belief that Ramsey thought truth played for logical consistency, of what might count as the best humanly attainable degrees of belief, and of the roles of short run and long run evaluations of systems of decision and degrees of belief. Adams applies the results of these investigations to the examples of the sort that have been used to motivate causal decision theory. He offers optimizing with respect to objective conditional probabilities as the appropriate "long run" evaluation of such cases, but he refrains from making any recommendation for the short term "single case" evaluation of them. The "single case" evaluation is made difficult by the agent's ignorance of the relevant objective conditional probabilities. It may be worth remarking that the evaluation Armendt recommends for the "single case" evaluation of these examples is an epistemic expectation of the objective conditional probability evaluation Adams recommends for their long run evaluation.

In the fourth paper, "Powers", Henry Kyburg goes along with the causal decision theorists in admitting that rational decision making appeals to a problematic causal notion, but he proposes to separate this notion from any application to acts construed as objects of scientific investigation. He is exploring a version of a tradition going back to Kant's treatment of the Freedom Antinomy, which distinguishes two aspects under which acts of an agent can be considered. When you consider your own acts as an agent the problematic notion of causal power applies, but scientific explanation and evidential relations to other events don't.[4] When these acts are construed (by someone else or by yourself) as events in the world they are open to scientific explanation and bear evidential relations to other events, but the problematic notion of causal power doesn't apply. Kyburg shows that his account will handle the examples originally used to motivate causal decision theory. Whether it can be extended in any natural way to handle cases of strategic reasoning in game theory (e.g. as discussed in Harper, this volume) where both sorts of reasoning seem to be appealed to in selecting a strategy, or cases (such as Sobel's [1986] popcorn example) where it's not just dominance but expectations of chances that seem to be relevant is a matter for further investigation.

PART II: RATIONAL BELIEF CHANGE

Peter Gärdenfors "Causation and the Dynamics of Belief" appeals to a model of belief contraction to illuminate a number of the puzzles that have plagued attempts to provide accounts of event causation. Isaac Levi (1976, 1980), William Harper (1978, 1979) and Gärdenfors (1979, 1980, 1982, 1986) have introduced models of rational belief that allow for hypothetical reasoning relative to weakenings or contractions that open the question with respect to a presently accepted proposition. Gärdenfors (1986) has developed this idea more extensively than anyone else, and he has applied it to illuminate explanation (1980). Here he applies it to good effect to illuminate event causation by getting clear about how beliefs about causation interact with various hypothetical revisions of belief states. Gärdenfors singles out the pragmatic constraints on which background assumptions get fixed in the relevant contractions as the critical determinate guiding reasoning about causal examples. He illustrates this by filling in appropriate details for some of the most puzzling examples in the literature. He also argues that Granger's (see Vol. I, pp. 3—21) account of causation is a special case of the general schema that guides Gärdenfors' contraction approach.

The relative immunity to revision that determines what gets preserved in Gärdenfors' contractions is closely related to Brian Skyrms' (1979) concept of resiliency. I think causal reasoning is just one of many areas where a notion like resiliency or relative importance among beliefs must be called upon to augment ideas based on high probability alone if we are to do justice to real life scientific reasoning. Gärdenfors puts qualitative constraints on contraction functions and specifies an interdefinability between contractions and revisions. In the next paper Spohn uses a sophisticated qualitative measure of relative importance to generate models of revision that meet qualitative constraints on revision that correspond to the Gärdenfors' constraints on contraction.

Wolfgang Spohn's "Ordinal Conditional Functions: A Dynamic Theory of Epistemic States" provides the most comprehensive dynamic model of qualitative belief states that has yet been developed. One of the salient difficulties facing any attempt to model rational belief change for common sense qualitative belief states is accommodating revision of previously accepted knowledge claims. This problem is especially acute for those who would model such states within a Bayesean framework where conditionalization plays a major role in the representation of rational belief change.

Once one has conditionalized on a fallible claim no later classical conditionalization can lead one to give it up. Strict Bayesean models, where fallible claims are never accepted and rational belief change is represented by Jeffrey conditionalization have a real advantage here. On these models any change introduced can be later revised. Isaac Levi (1976, 1979, 1980, 1983), Peter Gärdenfors (1979, 1981, 1982) and William Harper (1976, 1978, 1979) have all attempted to accommodate acceptance with possible later revision within a Bayesean framework. Levi and Gärdenfors explore rational belief changes which involve revisions of accepted claims, but neither has attempted to explicitly build into the representation of a belief state resources that would specify the outcomes of iterated sequences of revisions so as to rival the uniform treatment of belief change provided for by the strict Bayeseans. Harper attempted such a rival treatment by appealing to Popper functions (whose conditional probabilities are defined even for propositions having zero probability. Economists know similar functions as lexicographic-ordered probabilities). Just representing the belief state by a Popper function and rational belief change by conditionalization is not enough. One must also represent the output of the belief change as a new Popper function if iterated belief changes with revision of accepted claims are to be accommodated. In Harper 1976 conditional operators were used, while in Harper 1978 and 1979 the solution was to represent belief states as whole sets of hypothetical sequences of Popper functions. Spohn's ordinal conditional functions are a substantial improvement over any of these previous attempts. He starts with a grading of disbelief somewhat like Shackle's (1969) measure of potential surprize. This can also be regarded as a representation of relative firmness of accepted information much like that appealed to in Harper 1979 (pp. 454—456) to construct minimal revisions of bodies of information. Spohn uses ordinal valued functions to give great power to this representation of relative firmness, but his really innovative idea is an ingenious definition of conditionalization of ordinal conditional functions. Spohn's definition is well motivated by intuitions about minimal revision of representations of relative firmness and it has the great virtue of allowing for uniform treatment of very complex iterated belief changes.

Thanks to Spohn rational belief change for common sense qualitative belief states can be modelled with the same flexibility and power that Jeffrey rule changes offered to the strict Bayeseans. Spohn's model can be easily augmented with probabilities to provide a very rich Popper function representation that can cover all the changes modelled by the strict Bayeseans as well as the qualitative changes found in much common sense reasoning and in expert systems research. Among the interesting applications Spohn provides are quite sophisticated qualitative analogues to probabilistic accounts of independence and conditional independence.

PART III: STATISTICS

In "Four Themes in Statistic Explanation" Isaac Levi argues that cases in which statistical explanations might be considered causal are so because of a causal role played by an underlying deductive-nomological explanation. He makes interesting use of the Ehrenfest's (1959) classic book on the conceptual foundations of statistical mechanics to give non-trivial examples which illuminate causal themes in several types of statistical explanation. Levi offers a nice discussion of explanations of possibility in which his epistemic concept of serious possibility relative to an appropriate body of information plays a central role. He also provides an updated account of how his system with epistemic utilities and indeterminate credal states handles acceptance and direct inference. This discussion includes a serious objection to high probability acceptance rules and brings out the important difference between credal probability and confidence in Shackle's sense. This discussion complements the emphasis on relative importance of information in Gärdenfors' and Spohn's papers. It also provides good illustrations of the role of contraction in explicating the reasoning underlying causal themes in statistical explanation. Levi ends with a very nice discussion of the context dependency of acceptance which answers interesting questions raised by Tatiana Ehrenfest (1958) in the context of statistical mechanics.

Our last paper, Clark Glymour "Artificial Intelligence for Statistical and Causal Modelling" introduces the TETRAD program, a striking contribution Glymour and his associates have made to the problem of finding adequate causal models for non-experimental data. This is perhaps the most difficult problem many researchers in econometric modelling and other non-experimental behavioural sciences have to face. Standard data analytic techniques such as factor analysis and path analysis require assumptions about causal structure before they can be applied. These assumptions are usually linear causal models. Such models can be represented by directed graphs connecting some variables (the causes, independent or exogenous variables) with others (the effects, dependent or endogenous variables). The problem is due to the extremely large number of possible linear causal models corresponding to any given set of data variables. Glymour points out that for even a mere six variables there are 4^{15} possible models and that some econometric models have hundreds of variables. Normal practice is to appeal to the researchers sense of the situation being investigated to select some appropriate causal model. The problem is that such methods may well (indeed, almost certainly will) overlook some models that would better account for the data. The TETRAD program developed by Glymour and his associates is a sophisticated, computer search procedure to find improvements on any given linear causal model for a given body of data. Glymour provides a number of impressive examples where the TETRAD program has generated, from actual models proposed by behavioural scientists, new models that, by statistical tests standardly used, fit the data considerably better than the models proposed by the scientists.

Glymour introduces the TETRAD methodology with a fascinating account of the early work by Spearman and others on data analysis based on using certain equations (the tetrad equations) as constraints. This early work was superceded by what have become the main line data analytic techniques such as factor analysis and path analysis, but it used an idea about reasoning from data constraints to causal structure that Glymour and his associates were able to develop into the TETRAD methodology. The idea is that one looks for models in which the constraints are satisfied independently of any specific assumptions about the values of linear coefficients in the model. This idea of robustness of the constraints with respect to possible parameter values is closely related to what we have seen to be a recurrent theme in the papers in this volume — the theme that relative immunity to revision of information or what Skyrms has called resiliency must be appealed to in addition to high probability if causal reasoning is to be understood.

NOTES

[1] I believe David Lewis (1981) is responsible for coining this name.

[2] On this account the utility multiplier $u(A, B_i)$ represents the agent's conditional preference for A given state B_i, which may in some cases differ from $u(A + B_i)$ (see Armendt, this volume, pp. 13—14).

[3] This close interaction among these various problematic notions in the context of rational decision making can be exploited to bring intuitions about each into contact with intuitions about the others to afford opportunities for mutual clarification (see Harper, 1981, pp. 3—33), even if no one of these notions is as clear as we would like it to be.

[4] For Kant Freedom is the problematic notion and causality is part of the scientific picture.

REFERENCES

Bogdon, R. (1976), *Local Induction*, Dordrecht: D. Reidel.

Eells, E. (1982), *Rational Decision and Causality*, Cambridge: University Press.

Ehrenfest, P. and Ehrenfest, T. (1959), *The Conceptual Foundations of the Statistical Approach in Mechanics*, Translated by M. J. Moravscik, Cornell University Press.

Ehrenfest, T. (1958), 'On the Use of the Notion "Probability" in Physics', *American Journal of Physics* **26**, 388—392.

Fishburn, P. (1973), 'A Mixture-Set Axiomatization of Conditional Subjective Expected Utility', *Econometrica* **41**, 1—25.

Gärdenfors, P. (1979), 'Conditionals and Changes of Belief', *Acta Philosophica Fennica* **30**, 382—404.

Gärdenfors, P. (1980), 'A Pragmatic Approach to Explanations', *Philosophy of Science* **47**, 404—423.

Gärdenfors, P. (1981), 'An Epistemic Approach to Conditionals', *American Philosophical Quarterly* **18**, 203—211.

Gärdenfors, P. (1982), 'Rules for Rational Changes of Belief', in *Philosophical Essays Dedicated to Lennart Aqvist on his Fiftieth Birthday*, T. Pauli (ed.), Uppsala.

Gärdenfors, P. (1984), 'Epistemic Importance and Minimal Changes of Belief', *Australian Journal of Philosophy* **62**, 136—157.

Gärdenfors, P. (1986), 'The Dynamics of Belief: Contractions and Revisions of Probability Functions', *Topoi* **5**, 29—37.

Gibbard, A. and Harper, W. L. (1978), 'Conterfactuals and Two Kinds of Expected Utility', in Hooker, C. A., Leach, J. J., and McClennen, E. F. (eds.), Vol. 1, 125—162.

Harper, W. L. (1976), 'Rational Belief Change, Popper Functions and Counterfactuals', in Harper and Hooker (eds.), *Foundations of Probability Theory, Statistical Inference, and Statistical Theories of Science*, Vol. 1, 73—115. Dordrecht: D. Reidel.

Harper, W. L. (1976a), 'Ramsey Test Conditionals and Iterated Belief Change', also in Harper and Hooker (eds.), 117—136.

Harper, W. L. (1978), 'Bayesian Learning Models with Revision of Evidence', *Philosophia*, June 1978.

Harper, W. L. (1979), 'Conceptual Change, Incommensurability and Special Relativity Kinematics', *Acta Philosophica Fennica* **30**, 431—461.

Harper, W. L. (1980), 'A Sketch of Some Recent Developments in the Theory of Conditionals', in Harper *et al.* (eds.), 3—38.

Harper, W. L., Stalnaker, R., and Pearce, G. (eds.) (1980), *Ifs*, D. Reidel.

Herstein, I. and Milnor, J. (1953), 'An Axiomatic Approach to Measurable Utility', *Econometrica* **21**, 291—297.

Hooker, C. A., Leach, J. J., and McClennen, E. F. (1978), *Foundations and Applications of Decision Theory*, D. Reidel.

Jeffrey, R. C. (1965), *The Logic of Decision*, New York, 2nd edition (1983), University of Chicago Press.

Kahneman, D., Slovic, P., and Tversky, A. (eds.), *Judgment under Uncertainty: Heuristics and Biases*, Cambridge: University Press.

Kyburg, H. and Smokler, H. (eds.) (1964), *Studies in Subjective Probability*, New York: Wiley.

Levi, I. (1976), 'Acceptance Revisited', in Bogdon (1976), 1—71.
Levi, I. (1979), 'Abduction and the Demands for Information', Reprinted in Levi (1984), 87—106.
Levi, I. (1980), *The Enterprise of Knowledge*, Cambridge: M.I.T. Press.
Levi, I. (1983), 'Truth, Fallibility and the Growth of Knowledge', Reprinted in Levi (1984).
Levi, I. (1984), *Decisions and Revisions*, Cambridge: University Press.
Levi, I. (1986), *Hard Choices*, Cambridge: University Press.
Lewis, D. (1981), 'Causal Decision Theory', *Australasian Journal of Philosophy* **59**, 5—30.
Machina, M. (1982), '"Expected Utility" Analysis without the Independence Axiom', *Econometrica* **50**, 277—323.
Ramsey, F. (1926), 'Truth and Probability', Reprinted in Kyburg, H. and Smokler, H. (eds.) (1964).
Savage, L. J. (1954), *The Foundations of Statistics*, John Wiley and Sons Inc.
Shackle, G. L. S. (1969), *Decision, Order and Time in Human Affairs*, Cambridge: University Press, 2nd edition.
Skyrms, B. (1979), *Causal Necessity*, Yale University Press.
Skyrms, B. (1982), 'Causal Decision Theory', *Journal of Philosophy* **79**, 695—711.
Skyrms, B. (1984), *Pragmatics and Empiricism*, Yale University Press.
Sobel, H. (1978), 'Probability, Chance, and Choice: A Theory of Rational Agency', Manuscript.
Sobel, H. (1986), 'Notes on Decision Theory: Old Wine in New Bottles', *Australasian Journal of Philosophy* **64**, No. 4, 407—437.
Stalnaker, (1973), 'Letter to David Lewis', Reprinted in Harper *et al.* (1980).

WILLIAM HARPER
University of Western Ontario

PART I

DECISIONS AND GAMES

BRAD ARMENDT

CONDITIONAL PREFERENCE AND CAUSAL EXPECTED UTILITY

I am going to describe how we can make use of the idea of conditional preference to provide a foundation for Brian Skyrms' version of causal decision theory (*K*-expectation, or causal expected utility, decision theory). The foundation has the following two virtues: first, it puts the theory on an equal footing with competitors for which foundations have already been given; second, this foundation will provide the basis for a reply to the most serious objections made so far against causal decision theory, and against Skyrms' version of it in particular (more about this below).[1] I will only say a little about the other versions of causal decision theory developed by Gibbard and Harper (1976), Lewis (1981), and Sobel (1978). There are interesting differences between the theories, but in spirit they are very much alike. The foundation and the replies are immediate payoffs of using conditional preferences, but I think the treatment of conditional preference sketched below is also interesting for its own sake.

The foundation I will describe consists of (1) a set of axiomatic conditions on rational preference systems, and (2) the derivation of a representation theorem which shows that for any preference system satisfying the axioms there exist a probability measure P and utility functions U which represent the preferences, and which are related by the theory's general expected utility rule. The theorem has standard uniqueness results: for each preference system, P is uniquely determined and U is unique up to positive linear transformation. I should make it clear from the beginning that this foundation relies on formal results of Fishburn's (1973). My alterations of his formal theory are small. I do reinterpret his theory somewhat. The application to causal decision theory is new.

CAUSAL DECISION THEORY

Causal decision theories were proposed in reaction to the perceived failure of so-called *V*-maximization theories to adequately deal with problems like the Newcomb problem and the Twin Prisoners'

William L. Harper and Brian Skyrms (eds.), Causation in Decision, Belief Change, and Statistics, II, 3–24.

Dilemma[2]. In Richard Jeffrey's V-maximization theory the idea that actions are to be evaluated by their expected values takes the form

$$\text{(J)} \qquad V(A) = \sum_i P(C_i/A)\,V(A\ \&\ C_i),$$

for any A and any partition $\{C_1, \ldots, C_n\}$. This theory has the nice feature that no initial classification of acts, states, and consequences need be made: all appear as members of a Boolean algebra of propositions, and there is no requirement that acts and states of the world must be independent. So, though the notation suggests that the rule evaluates actions in terms of consequences, rule (J) is quite general, holding for any proposition A and any partition of C's. Iteration of (J) yields (J2) which might be useful to an agent who is better able to estimate the probabilities and values appearing in this rule than the ones appearing in (J):

$$\text{(J2)} \qquad V(A) = \sum_j P(B_j/A) \sum_i P(C_i/A\ \&\ B_j)\,V(A\ \&\ C_i\ \&\ B_j),$$

for any A and any partitions $\{B_1, \ldots, B_m\}$, $\{C_1, \ldots, C_n\}$. V-maximization is usually a good theory: most of the time an agent using V-maximization will accurately evaluate his alternatives. And I will show later that the utility rule given by the foundation I will present agrees with V-maximization in the many cases V-maximization gets right.

But (J2) can lead an agent astray in the problems we might call *causal counterexamples* to V-maximization. The best-known of these are the Newcomb problem and the Twin Prisoners' Dilemma, but here I will illustrate the problem for V-maximization with Fisher's smoking gene hypothesis.

Fisher's Smoking Gene Example

I believe that my disposition to smoke cigarettes and my disposition to contract cancer are genetically influenced by the same factor, which accounts for the statistical correlation between smoking and cancer. I believe that I cannot influence my genetic makeup, and that smoking itself is not harmful. I would enjoy smoking and attach small positive utility to the pleasure I would derive from smoking this cigarette. I attach large negative utility to contracting cancer. If I believe that the causal connections and statistical correlations are strong enough, and if I disvalue contracting cancer enough, compared to how much I value the pleasure of smoking, then (J) and *V*-maximization will lead me to the decision to refrain from smoking, in order to minimize the chances I have the gene and contract cancer.[3]

But this answer is wrong. If I believe (1) that nothing I do now can contribute to my having or not having the genetic factor, and (2) that the genetic factor is the main cause of my contracting cancer while smoking itself is *not* a cause of cancer, then the correlation between smoking and having the gene is *not* a good reason for avoiding the pleasure I would get from smoking. To act otherwise is, as Gibbard and Harper (1976) remark, to knowingly act so as to produce *evidence* for a strongly desired state of affairs (absence of the gene), without in any way *producing* the desired state, *even when such action has significant cost.*

An agent who follows *V*-maximization and the reasoning sketched above wrongly evaluates his alternatives because he fails to use his most specific and relevant information (or beliefs) in the given situation, namely his full information about the causal structure of his decision problem. This is so because (the agent believes that) his action in no way causes any of the states B_j to obtain, although the action is correlated with the states. In these situations, doing A_s (smoking the cigarette) is not a cause but is a symptom of B_g's obtaining (having the gene), so the conditional degree of belief $P(B_g/A_s)$ is greater than the belief $P(B_g)$. *V*-maximization leads the agent to act so as to raise (at significant cost) the *epistemic* probability of the desirable state (not having the gene) even though the agent believes such action cannot cause the state to occur. *V*-maximization recommends this because it ignores the information about the causal independence of the action and the state (no dependence from action to state), while attending to the information about the epistemic dependence of one on the other.

The various causal decision theories are designed to correct *V*-maximization by incorporating the agent's beliefs about the absence of a causal connection from his action to the states in his evaluation of his action. The basic idea shared by all the causal decision theories is that if the agent believes that the world may be in one of several states (or have one of several structures) whose occurrence are not causally influenced by his action, and which each affect the chances the consequences have of being caused by the action (or which each affect the values of the consequences), then he should evaluate his action this way: For each of the possible states or structures, find the value the action has if that state holds; then find the *overall* value of the action by taking a weighted average of these values, using as weights the degrees of belief in each state's being the actual one. The theories differ in their description of the states and structures to be considered and in their analysis of the action's value for each state. The theories of Gibbard and Harper (1976) and Lewis (1981) both recommend that the agent consider counterfactual conditionals describing possible causal patterns the world might have that are relevant to the actions and consequences in question.[4] Which of these causal patterns obtains is taken to be outside the agent's control, and in both theories the agent is told to weight the values he gives to the possible consequences by his degrees of belief in the competing causal counterfactual conditionals. The appropriate conditionals are *causal* in the sense that in the smoking gene problem, for example, the agent would be expected to assign a substantial degree of belief to "I smoke $\Box\!\!\rightarrow$ I enjoy my cigarette," but a very low degree of belief to "I smoke $\Box\!\!\rightarrow$ I get cancer," given that he believes that his smoking does not cause him to have the gene, and that the only causal connection between the smoking and getting cancer is through this genetic factor. The appropriate causal counterfactuals are not "backtracking counterfactuals."

That is all I will say here about the other versions of causal decision theory. Skyrms' version is not formulated in terms of counterfactual conditionals, and it is to his theory that I will now turn.

Consider the smoking gene example and recall that the V-maximization formula (J2) does not incorporate a correct summary of my beliefs about the situation. The $P(B_j/A)$'s which appear in (J2) will reflect my belief that smoking is correlated with having the gene, but not my belief that smoking does not cause me to have it. We might delete the misleading influence that the correlation contributes to the evaluation of act A by replacing these conditional probabilities with the simple unconditional degrees of belief $P(B_j)$'s. If we take note of this alteration of our expected utility rule by writing the utility function U, and we note the special requirement on the states B_j by writing them as K_j, (J2) then becomes

$$\text{(S)} \qquad U(A) = \sum_j P(K_j) \sum_i P(C_i/A \;\&\; K_j)\, U(A \;\&\; C_i \;\&\; K_j).$$

(S) will agree with (J2) and V-maximization whenever the states K

which influence the chances or utilities of the consequences are statistically independent of the agent's choice of action. But when the states and the choice are not statistically independent, (J2) and (S) will very likely give different values to A, and these differences may lead to different recommendations when the agent chooses the alternative with the highest utility. It is important to note that our justification for suggesting (S) is that the states K_j are believed *causally relevant* to the action A's production of its possible consequences C_i and they are believed *causally independent* of the action (no dependence from action to state). If the K_j's satisfy those conditions, but not otherwise, then we correctly summarize the agent's beliefs by using (S). Skyrms calls partitions which in a given decision situation describe the causally relevant (possible) states of the world which are outside the agent's influence *K-partitions*. Skyrms' version of causal decision theory recommends that the agent choose the act which has maximum causal expected utility, or K-expected utility, given by (S).[5]

OBJECTIONS TO CAUSAL EXPECTED UTILITY THEORY

There has been considerable debate about whether or not we need to adopt causal decision theory in order to make the best choices in problems like these. Sophisticated V-maximization theories have been developed, and their defenders have argued that causal decision theory is superfluous. This view is mistaken, but I will not discuss it here.[6] Other criticisms have been made of Skyrms' causal expected utility theory that I *shall* consider.

It is easy to see that the key to applying the theory in the situations where it is most needed is the identification of an appropriate K-partition. An agent needs to find a set of propositions $\{K_j\}$ which are such that (a) their use in (S) is correct for the problem confronting him; that is, when they are used in (S) his evaluation of the action incorporates all his information/beliefs relevant to the problem; and (b) they are practical – the agent needs to have some idea what the values of the degrees of belief and utilities appearing in (S) actually are. As I said above, Skyrms (1979) has described the K-partition appropriate to a given problem as a partition of maximally specific descriptions of the factors outside the agent's influence which are causally relevant to the outcomes (that he cares about) of the alternatives available to him.

Now the theory has been criticized on the grounds that the appropriate *K*-partitions for decision problems must be carefully selected (which is true), that general use of the theory seems to depend upon a general way of finding an appropriate *K*-partition, and that this task requires an understanding on the agent's part of the relations "state *S* is outside my influence" and "state *S* is causally relevant to act *A*'s having outcome *C*" that is more subtle than is reasonable to require for a useful decision theory, even a normative one. Further, it has been said that if Skyrms' theory lacks a representation theorem, it lacks a theoretical guarantee that it is broadly applicable – a guarantee possessed by other decision theories, including *V*-maximization.[7] The axiomatic system and representation theorem I shall describe provide a direct response to the latter complaint about the theory. We shall also find an answer to the former objections concerning the problem of the selection of *K*-partitions.

It is worth pointing out that for an agent with a given decision problem the task of selecting an appropriate *K*-partition is an empirical one. And a *K*-partition is correct for a given decision problem by being correct for the agent's *beliefs* and *preferences* about the problem. The causal decision theories are more successful than *V*-maximization by better incorporating the agent's beliefs, particularly his causal beliefs, about his problem into his act evaluations – the Newcomb game and the smoking gene example would be no problems at all for *V*-maximization if the agent did not have the beliefs ascribed to him in those situations. This is not a point on which the critics of causal expected utility theory go wrong, but it is worth mentioning because it leads to the idea that we should look to the agent's preference system, which underlies his beliefs and desires, if we want to describe the selection of appropriate *K*-partitions for his decision problem. In what follows, I will be able to state sufficient conditions for the propriety of *K*-partitions in terms of their behavior in the agent's preference system. And it will turn out that these conditions correspond quite well with Skyrms' description given above.

PREFERENCE AND CONDITIONAL PREFERENCE

The descriptions of the entities for which our agent is supposed to have preferences[8] are given, first, by propositions which may describe acts, states, or consequences, and second, by mixtures of propositions. So the agent may have in his system preferences for going swimming, for its raining, for catching a cold, and so on. These preferences are represented in our system by corresponding propositions "I go swimming," "It will rain tomorrow," "I catch a cold," and so on. In addition, mixtures or gambles on propositions will be taken to be descriptions of objects of preference. So also appearing in our system will be mixtures like $0.6(R) + 0.4(-R)$, where R is "It will rain tomorrow;" and $0.2(S \& C) + 0.5(S \& -C) + 0.3(-S)$, where S is "I go swimming," and C is "I catch a cold." Such mixtures can be thought of as lotteries, where the mixing coefficients give the odds on each of the possible outcomes. This system will assume that the agent has a rich set of these preferences for mixtures (including mixtures of mixtures, and so on). They form a *mixture set*, almost in the sense of Herstein and Milnor (1953): the exception is that here mixing is restricted to sets of propositions which form partitions, rather than being closed for all sets of propositions whatsoever. (This differs from Herstein & Milnor, who take the basic elements of mixture sets to be incompatible payoffs; there is no need in their system to further specify that mixtures be on incompatible outcomes). This modification of Herstein–Milnor's (and Fishburn's) theory will be discussed further below. Two points that I want to bring out here are that (1) Fishburn's theory and my adaptation of it follow, as do many other decision theories, the von Neumann and Morgenstern approach to utility in that the preference systems are taken to include mixtures or lotteries;[9] (2) this theory deviates from Fishburn's and most other theories, however, in its use of *propositions* and mixtures of *propositions*, rather than the more common practice of interpreting preferences as preferences for nonpropositional acts, either primitive or regarded, for example, as functions from states to consequences; in this respect it follows the Jeffrey–Bolker "monoset theory".

The set of preferences so far described, and indeed the full set of preferences described below, are assumed to be ordered by a relation $\geqslant$, interpreted as "is at least as preferable as." The entire system of ordered (rational) preferences will be assumed to satisfy a set of axiomatic conditions that I will present below; notice that the ordering assumption includes requirements that preference is transitive, and that any pair of elements of the ordering are comparable.

Now the system described so far is further enriched by the addition of *conditional preferences*. The agent may entertain an hypothesis about the world, and under that hypothesis he may find that his preference-attitudes toward various states, acts, etc. differ from the preference-attitudes he directs toward those states, acts, etc. under other hypotheses, or under the trivial hypothesis *T* (which, for convenience, is how I will regard unconditional preferences). That is, under one hypothesis *H* the agent may rank his preferences (conditional on *H*) quite differently from the ranking of his unconditional preferences, or the ranking of his preferences conditional on hypotheses *J*, *K*, ... Of course, conditional hypotheses need not *always* alter the ranking of preferences. For example, under my hypothesis that it rains here this afternoon, my preference for swimming this afternoon at the local beach is ranked below my unconditional preference for swimming this afternoon at that beach. But under the hypothesis that it snows today in Tibet, my preference for swimming this afternoon is not perturbed.

I shall write *P*,*Q* and *m*,*Q* for the conditional preferences for *P* and for *m*, under the hypothesis that *Q*. As I noted just above, I shall regard unconditional preferences as preferences conditional on the tautology *T*, and sometimes will write them *P*,*T*. A conditional preference, I emphasize, is not a preference for a conditional proposition. And *P*,*Q* is to be understood as the agent's *present* preference for *P*, under his hypothesis that *Q* – not (necessarily) as the preference he *would have* for *P* if *Q* were true, or if he were to *believe* *Q* true.

Sometimes a conditional preference may have the same ranking as an unconditional preference for a corresponding conjunction. The swimming examples above may well be illustrations of this; my preference for swimming, under my hypothesis that it rains, may be ranked the same as my unconditional preference for the conjunction of swimming and its raining. And similarly for my preference under my hypothesis about the snow in Tibet. But this is not always the case. Consider the preferences *M*,*H* and *H*,*M* where *M* is "I have medical insurance," and *H* is "I am hospitalized for a serious illness." My preference for having medical insurance, under the hypothesis that I am hospitalized, is considerably greater than my preference for being hospitalized, under the hypothesis that I have insurance. Of course, if both preferences were ranked with the corresponding conjunction, then I would be indifferent between them. So we will not assume that conditional preferences are always ranked with corresponding conjunctions, though in many cases it may turn out that they are.

What about preferences of the forms P,P and $-P,P$? I regard such conditional preferences as well-formed and non-trivial; they play a significant role in the description of the proper use of the causal expected utility rule (S). Under his hypothesis that the actual world is a P-world, the agent has preferences for its being a P-world and for its being a $(-P)$-world. It is important to keep in mind when considering $-P,P$ that this is not a preference for $-P$ & P, and it is certainly not a belief in $-P$ & P. One may have a well-defined unconditional preference for a proposition one strongly believes false, and one may have a conditional preference for a proposition assumed false – not just because the proposition may not be assumed to be known false, but mainly because preference and desire may be directed toward propositions in which one does not believe. I may desire that Pegasus be alive and willing to carry me wherever I please; and I may still desire this, under my hypothesis that Pegasus never lived in the actual world.

It will often turn out that making the hypothesis P (or the hypothesis $-P$) does *not* perturb an agent's preference for P. Let W be "Candidate C wins the election," and let S be "I go swimming this afternoon;" it is quite plausible that I be indifferent between W, W,W, and $W,-W$; and also between S, S,S, and $S,-S$. But consider the Newcomb problem, letting A_2 be "I take both boxes," and the smoking gene problem, letting A_s be "I smoke the cigarette." In the former case, it is highly plausible that A_2,A_2 is ranked below A_2, and $A_2,-A_2$ is ranked above A_2. Under the hypothesis that I take both boxes, my preference for taking both boxes is diminished, since worlds in which I take both boxes are worlds where an empty opaque box is likely. And since in worlds in which I do not take both boxes a filled opaque box is likely, the hypothesis that I do not take both raises my preference for taking them. For similar reasons, in the smoking gene example I prefer $A_s,-A_s$ to A_s to A_s,A_s. Another example that reinforces the idea that preferences like A,A and $A,-A$ make nontrivial sense is the story of the man who met death in Damascus, from Gibbard and Harper (1976): when the story is elaborated in the way Gibbard and Harper do, it is clear that the man prefers $A,-A$ to A to A,A, where A is "I go to Aleppo." In the theory that follows we will see that conditional preferences like A,A and $A,-A$ are important. The failure of an act proposition A to be ranked equally to A,A is an indication that use of the causal expected utility rule (S), rather than a simpler special case of it, is appropriate in evaluating A.

Two further points about conditional preference: First, though the examples so far mentioned have all been of the form P,Q where P is a proposition, conditional preferences for mixtures of the form m,Q are also included in the agent's preference system. But *hypotheses* are always taken to be propositions rather than mixtures; under a particular hypothesis H the agent's preferences form a mixture set (in our sense, mixtures only on partitions) written $\mathbf{M_H}$. Second, notice that mixtures are only formed under single conditional hypotheses; the agent is *not* assumed to have preferences like $0.6(P,Q) + 0.4(-P,R)$. Luce and Krantz (1971) have a well-known theory in which disjunctive preferences, each disjunct conditional on different hypotheses, are assumed to be in the preference system. And Balch and Fishburn (1974) present such a theory with mixtures like $0.6(P,Q) + 0.4(-P,R)$. I regard such preferences as intuitively very problematic and formally unnecessary.

The remarks made so far clearly amount to less than a complete account of conditional preference, but at this point I shall proceed with the description of the theory's preference systems and representation theorem. I claim that the assumptions made by this theory about conditional preference are actually not very strong, and that whatever correct account of conditional preference emerges from the intuitions I have appealed to will be consistent with those assumptions.[10]

DERIVATION OF CAUSAL EXPECTED UTILITY THEORY

I will discuss the axioms for rational preference systems and the representation theorem soon. Let us look ahead in this section and see how the results stated there provide a foundation for causal expected utility decision theory. Fishburn's 1973 representation theorem shows that a preference system $\langle \mathbf{X}, \geqslant \rangle$ which satisfies the axioms to be stated below can be represented by a probability measure P and utility function U. U is linear and order-preserving, P is unique, U is unique up to positive linear transformation, and the following utility rule holds for all x in $\mathbf{X}$:

$$\text{(F1)} \qquad U(x,B \vee C) = P(B/B \vee C)\, U(x,B) + P(C/B \vee C)\, U(x,C)$$

whenever x,B and x,C are in $\mathbf{X}$, and B and C are incompatible.[11] And for A and a partition of propositions $B_1, B_2, \ldots, B_n$ such that A,B_i is in **X** for each i, iterated application of (F1) gives us the rule

$$\text{(F2)} \qquad U(A) = \sum_i P(B_i)\, U(A,B_i).$$

If U is a utility function and P the probability measure which represent my preferences, (F2) says that, for example, the utility I attach to swimming this afternoon $U(S)$ is equal to my degree of belief in rain $P(R)$ times the utility of swimming under the hypothesis that it rains, plus $P(-R)$ times the utility of S under the hypothesis $-R$. In the smoking gene problem, the utility I attach to smoking, A_s, is equal to my degree of belief that I have the gene, $P(G)$, times the utility of A_s under the hypothesis G, plus $P(-G)$ times $U(A_s,-G)$. This agrees with causal decision theory's evaluation of A_s, since the weights are *not* $P(G/A_s)$ and $P(-G/A_s)$ as in V-maximization theory.

When we compare (F2) with the V-maximization rule (J) it is clear that they agree when for each i, $P(B_i) = P(B_i/A)$ and $U(A,B_i) = U(A \;\&\; B_i)$. Since the function U is order-preserving, the latter conditions hold when I am indifferent between A,B_i and $A \;\&\; B_i$. So if I believe the states B_i are statistically independent of A, and my conditional preferences for A under the hypotheses B_i are the same as my unconditional preferences for the conjunctions $A \;\&\; B_i$, then evaluation of A using V-maximization will agree with this theory's evaluation of A.

If A is an act proposition, however, we are likely to want to evaluate it in terms of its possible *consequences*, rather than to use some other partition of states. And in any interesting decision problem, of course, the statistical independence will *not* hold – I will believe that A's possible consequences *are* influenced by whether or not I do A. There is another direction we may take, though, in seeking cases where the two rules agree: If I am indifferent between A and A,A (doing A, under the hypothesis I do A), and if for all i, I am indifferent between $A \;\&\; C_i$ and $A,(A \;\&\; C_i)$, then it is easy to see that (F2) agrees with rule (J). By order-preservation of U,

$$U(A) = U(A,A) = \sum_i P(A \;\&\; C_i/A)\, U(A,A \;\&\; C_i).$$

So by substitution,

$$= \sum_i P(C_i/A)\, U(A \;\&\; C_i).$$

So in such cases I may use the V-maximization rule (J) and I will get the correct value for $U(A)$. I think it turns out that these conditions, that $A \sim A,A$ and that A & $C_i \sim A,(A$ & $C_i)$, are satisfied in the ordinary decision situations where we expect V-maximization to give the correct evaluations of the available acts. The first says that my hypothesis that I do A does not perturb my preference for A, and it was discussed above. The second says that my preference for A, under my hypothesis that I do A and consequence C_i results, is equal to my unconditional preference for the conjunction of A and result C_i. In ordinary situations I think this simply requires that the C_i's be accurate descriptions of the possible consequences of doing A that I care about. Call it a requirement that the C_i's are *well-selected* consequence descriptions.

When does (F2) take the form of rule (S), the causal expected utility rule? Well, it does if it agrees with V-maximization, and if V-maximization in turn agrees with causal expected utility theory (i.e. with (S) for some appropriate K-partition). So in cases where that is true and where $A \sim A,A$, (F2) yields the causal expected utility rule which agrees (as it should) with V-maximization.

What about cases in which $A \nsim A,A$? Such cases include, as I have claimed above, the causal counterexamples to V-maximization. In these cases A is correctly evaluated by the causal expected utility rule when appropriate K-partitions can be found. How does (F2) yield the rule (S)? If we suppose that the agent is *not* indifferent between A and A,A, then for most state descriptions B_j we have little reason to expect that he will be indifferent between A,B_j and $A,(A$ & $B_j)$. But *some* state descriptions B_j may align those preferences after all: among the descriptions which do this are descriptions which make up appropriate K-partitions. For example, in the smoking gene story my hypothesis that I smoke the cigarette A_s devalues my preference for A_s. But consider my preferences A_s,G and $A_s,(A_s$ & $G)$ where G is "I have the gene." My preference for A_s, under my hypothesis that I have the gene, is *not* devalued by the additional hypothesis that I smoke. My preference for A_s was already devalued by the hypothesis G and (given the story assumed in this example) is not further affected by supposing that I do light up. Similarly for $-G$. Having or not having the gene is what matters in this decision problem, and once I suppose that I do or that I do not, my preference for A_s is fixed, with respect to the additional hypothesis A_s. This is true in general of propositions which are members of adequate K-partitions – the hypothesis that one of them holds in the actual world fixes the value of the act to the extent that whether or not the additional hypothesis that the act is done is added, the agent's preference for it remains the same.

So if we return to (F2), assume that $A,B_j \sim A,(A \ \& \ B_j)$, and then, in recognition of the special character of this partition of state descriptions, rewrite them as K_j's, we have

$$U(A) = \sum_j P(K_j)\, U(A,A \ \& \ K_j).$$

This will generate the rule (S) if we can find a partition of consequence descriptions which for each K_j satisfies our condition stated earlier for well-selected C_i's: If there is a partition $C_1, \ldots, C_n$ such that $A,(A \ \& \ C_i \ \& \ K_j) \sim A \ \& \ C_i \ \& \ K_j$ for all i,j, then we can use (F2) to analyze $U(A,A \ \& \ K_j)$ in terms of the possible consequences:

$$\begin{aligned} U(A,A \ \& \ K_j) &= \sum_i P(C_i/A \ \& \ K_j)\, U(A,A \ \& \ C_i \ \& \ K_j) \\ &= \sum_i P(C_i/A \ \& \ K_j)\, U(A \ \& \ C_i \ \& \ K_j). \end{aligned}$$

And so by substitution into the first equation of this paragraph:

$$U(A) = \sum_j P(K_j) \sum_i P(C_i/A \ \& \ K_j)\, U(A \ \& \ C_i \ \& \ K_j).$$

This is the causal expected utility rule (S).

So when $A \not\sim A,A$ our theory endorses rule (S) for partitions $K_1, \ldots, K_m$ and $C_1, \ldots, C_n$ which satisfy

$$\begin{gathered} A,K_j \sim A,(A \ \& \ K_j) \quad \text{and} \\ A,(A \ \& \ C_i \ \& \ K_j) \sim A \ \& \ C_i \ \& \ K_j, \end{gathered}$$

for all i and j. This is a nice result since these are exactly the conditions that we should expect appropriate K and C partitions to satisfy. Skyrms describes the appropriate K-propositions as maximally specific descriptions of the factors the agent believes are outside his influence and causally relevant to those outcomes of his available alternatives which matter to him. When such a description K_j is made a conditional hypothesis, it may of course perturb the agent's preference for one of his alternatives A: $A_s,G > A_s$ in the smoking gene example. But once that hypothesis is made, the uncertain states of the world which may influence the outcome of the action (including the states which are correlated with but not caused by the action) are fixed, and the additional hypothesis that A is done should not further perturb the preference for A. The condition $A,K_j \sim A,(A \,\&\, K_j)$ captures the idea behind Skyrms' appropriate K-partitions, and it does so in the desirable way mentioned earlier: adequate K-partitions are picked out by reference to the way they behave in the agent's preference system.

Now any foundation for rational decision theory, even one which provides the nice results shown above, depends much for its adequacy on the assumptions about preference contained in its axioms. In the next section, I shall present the axioms for rational preference and the theorem. Discussion of the adequacy of the axioms can be found in Armendt (1983) and (1986a). I shall conclude with some sketchy remarks about the possibility of providing a different foundation – one more like the Jeffrey–Bolker foundation for V-maximization.

THE AXIOMS AND THEOREM

The structure which is interpreted as the agent's preference ranking is a collection of mixture sets each of whose basic elements are members of a Boolean algebra of propositions. The mixture sets are like Herstein–Milnor mixture sets, except that only mixtures of partitions are defined. Each mixture set corresponds to the agent's preferences under the hypothesis that some one element of $\mathscr{E}$ is true. We start with our set of propositions $\mathscr{E}$ that describe states, acts, and consequences. After deleting the contradictory proposition, we construct a set **M** of all mixtures or gambles on partitions of propositions in $\mathscr{E}'$ ($\mathscr{E}' = \mathscr{E} - \phi$). **M** is a large set containing all such mixtures, and $\mathbf{M} \times T$ is the set of all the agent's unconditional preferences.

A1. $\mathscr{E}$ is a Boolean algebra of propositions. And $\mathscr{E}' = \mathscr{E} - \phi$.

A2. **M** is a mixture set formed from $\mathscr{E}'$, i.e. $\mathscr{E}' \subset \mathbf{M}$, and:

a. *Closure under mixing*: For all $A,B \in \mathscr{E}'$ such that $A \,\&\, B = \phi$, and all $\alpha \in [0,1]$,

$$\alpha A + (1 - \alpha)B \in M;$$

and for all mixtures m on $A_1, A_2, \ldots, A_n$ and for all B disjoint from each A_i:

$$\alpha m + (1 - \alpha)A_i \text{ and } \alpha m + (1 - \alpha)B \text{ are in } \mathbf{M};$$

b. *Identity*: for all m_1, m_2 in **M**, and for all $\alpha,\beta \in [0,1]$:

$$1(m_1) + 0(m_2) = m_1$$
$$\alpha m_1 + (1 - \alpha)m_2 = (1 - \alpha)m_2 + \alpha m_1$$
$$\beta[\alpha m_1 + (1 - \alpha)m_2] + (1 - \beta)m_2 = \alpha\beta m_1 + (1 - \alpha\beta)m_2.$$

The set **X** is the set which contains *all* the agent's preferences, conditional and unconditional. It contains all the elements of $\mathbf{M} \times T$ together with the elements of $\mathbf{M_P}$ for every proposition P in $\mathscr{E}'$. For each P in $\mathscr{E}'$, $\mathbf{M_P}$ is interpreted as the set of the agent's preferences for the elements of **M** under the hypothesis P. It is a (possibly improper) subset of $\mathbf{M} \times P$. It is assumed to be a mixture set in the sense given above. **X** is the union of all the $\mathbf{M_P}$'s and is ordered by the relation $\gtrsim$. Notice that mixtures of conditional preferences are only allowed when they are formed under the same hypothesis: **X** contains $[\alpha P + (1 - \alpha)Q],R$ but not $\alpha P,R + (1 - \alpha)Q,S$.

A3. For all $P \in \mathscr{E}'$, $\mathbf{M_P} \subseteq \mathbf{M} \times P$ is a mixture set (see A2). Also, $\mathbf{M_T} = \mathbf{M} \times T$.

A4. The set **X** is the union of all the $\mathbf{M_P}$'s and is ordered by $\gtrsim$.

We now come to a pair of axioms which are Herstein–Milnor's axioms for mixture set functions generalized to **X** (which is a union of mixture sets), and incorporating our restriction that mixtures be on disjoint propositions

A5. For all A, $B \in \mathscr{E}'$, all x,A, $y,A \in \mathbf{M_A}$ such that mixtures on x,A and y,A are defined, and all $z,B \in \mathbf{M_B}$, these sets are closed in $[0,1]$:

$$\{\alpha: (\alpha x + (1 - \alpha)y),A \gtrsim z,B\} \quad \text{and}$$
$$\{\alpha: z,B \gtrsim (\alpha x + (1 - \alpha)y),A\}.$$

A6. For all A, $B \in \mathscr{E}'$, all x,A, $y,A \in \mathbf{M_A}$, and for all z,B, $w,B \in \mathbf{M_B}$, if $x,A \sim z,B$ and $y,A \sim w,B$, and if the mixtures below are defined, then

$$[\tfrac{1}{2}x + \tfrac{1}{2}y],A \sim [\tfrac{1}{2}z + \tfrac{1}{2}w],B.$$

Setting $A = B$, axioms A5 and A6 are almost the Herstein–Milnor axioms, and in the absence of any restrictions on mixing, would imply that Herstein–Milnor's axioms are satisfied by the elements of mixture set $\mathbf{M_A}$ for every A. So then (via the Herstein–Milnor theorem) for every $\mathbf{M_A}$ there exists a real-valued utility function U_A which is linear and order-preserving over $\mathbf{M_A}$, and is unique up to positive linear transformation. The next axiom, A7, yields these results even in the presence of the restriction that only mixtures on partitions are defined:

A7. For all $A \in \mathscr{E}'$, for all $x,A \geqslant y,A$, there exist disjoint P and Q such that P,A, $Q,A \in \mathbf{M_A}$ and

$$P,A \geqslant x,A \geqslant y,A \geqslant Q,A.$$

Axiom 8 guarantees that whenever $A \,\&\, B = \phi$ and x,A and x,B exist, then so does $x,(A \vee B)$. Axiom 9 then states Fishburn's averaging principle. Axiom 10 is a non-triviality condition.

A8. For all A, $B \in \mathscr{E}'$ such that $A \,\&\, B = \phi$, and for all $x \in \mathbf{M}$, if $x,A \in \mathbf{M_A}$ and $x,B \in \mathbf{M_B}$, then $x,(A \vee B) \in \mathbf{M_{A \vee B}}$.

A9. For all x,A and x,B in X, if $A \,\&\, B = \phi$ and $x,A \geqslant x,B$, then $x,A \geqslant x,(A \vee B) \geqslant x,B$.

A10. For some x, $y \in \mathbf{X}$, $x > y$.

Axiom 11 is required to generate the comprehensive utility function U on $\mathbf{X}$ from the many U_P's on the $\mathbf{M_P}$'s. From the x and y whose existence it asserts, a gamble z on x and y (or on stand-ins for which the gamble is defined) will be found such that $z,A \sim z,B$. This axiom denies that there are two incompatible hypotheses A and B such that every element of $\mathbf{M_A}$ is preferred to the corresponding element of $\mathbf{M_B}$, and it is a fairly strong structural condition:

A11. For all A, $B \in \mathscr{E}'$ such that $A \,\&\, B = \phi$, there exist x, $y \in \mathbf{M}$ such that $x,A > x,B$ and $y,B > y,A$.

Axioms A1–A11 imply the existence of a utility function U on $\mathbf{X}$ which is linear, order-preserving, and unique up to positive linear transformation, and also the existence for each incompatible A and B in $\mathscr{E}'$ of unique non-negative real numbers which sum to 1, $P_{A \vee B}(A)$ and $P_{A \vee B}(B)$, such that

$$U(x,A \vee B) = P_{A \vee B}(A)\, U(x,A) + P_{A \vee B}(B)\, U(x,B),$$

for all x,A, $x,B \in \mathbf{X}$ (Fishburn, Th. 3). Axiom 12 is required to guarantee the additivity of the probabilities P_A and the chain condition: $P_C(A) = P_C(B)\, P_B(A)$, whenever $A \Rightarrow B \Rightarrow C$:

A12. For all $A,B,C \in \mathscr{E}'$ that are pairwise incompatible, if there is an $x \in \mathbf{M}$ such that $x,A \sim x,B$, then there is a $y \in \mathbf{M}$ such that exactly two of y,A, y,B, and y,C are indifferent.

Axioms A1–A12 imply the representation theorem:

THEOREM (Fishburn, 1973). *If $\mathscr{E}$, $\mathbf{X}$, and $\geqslant$ satisfy axioms* 1–12 *above, then there is a real-valued function U on* $\mathbf{X}$ *and a finitely-additive probability measure P_A on $\{A \,\&\, B : B \in \mathscr{E}\}$ for each $A \in \mathscr{E}'$ such that:*

i. *$x,A > y,B$ iff $U(x,A) > U(y,B)$, for all x,A and y,B in* $\mathbf{X}$;
ii. *$U(\cdot,A)$ is linear (as a function on $\mathbf{M_A}$) for each A in* $\mathscr{E}'$;
iii. *$P_C(A) = P_C(B)\,P_B(A)$ whenever $A \Rightarrow B \Rightarrow C$, $A \in \mathscr{E}$, and $B, C \in \mathscr{E}'$*;
iv. *$U(x,A \vee B) = P_{A \vee B}(A)\,U(x,A) + P_{A \vee B}(B)\,U(x,B)$ whenever x,A, $x,B \in \mathbf{X}$ and $A \,\&\, B = \phi$*;

furthermore, the P_A's are unique and U is unique up to positive linear transformation.

Clause (iii) of this theorem provides measures P_B even when $P_T(B) = 0$. Note that whenever $A \Rightarrow B$ and $P_T(B) > 0$, $P_B(A) = P_T(A/B)$. Clause (iv) of the theorem is the general form of our theory's rule (F1) relating the utilities of a proposition or mixture under different conditional hypotheses.

Proof: Fishburn's system does not include the restrictions that appear in A2 (a), the closure condition on mixing, which restrict mixing to *incompatible* propositions. (This restriction permits us to regard mixtures as always on partitions, since we can think of $\alpha P + (1 - \alpha)Q$ as $\alpha P + (1 - \alpha)Q + 0[-P \,\&\, -Q]$. Fishburn does not include such a restriction because he takes acts, rather than propositions, to be the object of preference.) Restricting the closure condition on mixing in this way leads to the introduction of Axiom 7, which is not Fishburn's, into our system. The proof of the theorem for our system is as in Fishburn (1973), with the following modifications:

First, one can check the proof of the Herstein–Milnor theorem to see that it applies to the sort of mixture sets we employ (mixture sets such that the closure condition on mixing asserts only the existence of mixtures on certain subsets of elements in the set, i.e. disjoint propositions and mixtures of them): the checking of the proof is completely straightforward, up to the assertion that is Theorem 7, and the subsequent definition of the utility u, in the original Herstein–Milnor paper (1953). These concluding steps of the Herstein–Milnor theorem require that for arbitrary r_0 and r_1 there are elements a and b such that r_0 and r_1 fall in S_{ab} (the subset of the ordered preferences lying between elements a and b), and such that mixtures on a and b are defined. The axiom that has been added, Axiom 7, guarantees that such disjoint elements a and b exist. The proof then goes through as in the original Herstein–Milnor paper.

So alter Fishburn's proof as follows: Where the antecedents of assertions in Fishburn's proof mention mixtures, add to the antecedents of the assertions that the mixtures are well-defined. Where the consequents of assertions in Fishburn's proof assert the existence of mixtures (for scaling the utility functions) on x and y, say $\alpha x + (1 - \alpha)y$, substitute assertion of the existence of $\alpha x^* + (1 - \alpha)y^*$, which is obtained as follows:

Suppose $x \geqslant y$. Apply Axiom 7 to find disjoint P and Q such that $P \geqslant x \geqslant y \geqslant Q$. Use the version of Herstein–Milnor that holds for the sets of mixtures on partitions to obtain

$$x^* = \beta_x P + (1 - \beta_x)Q \sim x, \quad \text{and}$$
$$y^* = \beta_y P + (1 - \beta_y)Q \sim y.$$

So $\alpha x^* + (1 - \alpha)y^*$ is well-defined (Axiom 2a), and is ranked with $\alpha x + (1 - \alpha)y$ (Herstein–Milnor for mixtures on partitions). The mixture $\alpha x^* + (1 - \alpha)y^*$ will play the role in scaling our system's utilities that the mixture $\alpha x + (1 - \alpha)y$ (which may not be well-defined) played in Fishburn's proof.

COMPARISON WITH JEFFREY–BOLKER; SPECULATION ABOUT ANOTHER FOUNDATION

The foundation presented above is, in some important ways, better than others in the literature: most important, it is a foundation for the best theory of rational choice available, and it generates that theory in a natural way. In other respects, this foundation fares no worse than others: the constraints on rational preference systems imposed by the axioms are at least as plausible as conditions for rational preference as are those of other theories; the richness assumptions about the preference orderings are strong, but no stronger than assumptions generally made by other theories. In one respect, however, another theory has an advantage this theory lacks: the Jeffrey–Bolker theory has, as I noted before, the very nice feature that lotteries or mixtures are not built into the preference ordering – so in that theory it is really true that probability and utility measures are derived from purely *qualitative* preference.[12] It is true that the Jeffrey–Bolker foundation also has unusual uniqueness results, weaker than those of other theories, but in my view this does not detract from the elegance of the theory. In the terminology of Krantz *et al.* (1971), the Jeffrey–Bolker theory provides a *fundamental measurement* of preference by probability and utility, whereas other theories do not.

Can a theory with this virtue of the Jeffrey–Bolker theory be given for conditional preference, generating causal decision theory in the natural way I have described above? I think the answer is probably "yes", though I do not have the theory now. I cannot say very much more than it is a project worth pursuing, but the following seems worth mentioning:

1. Since (a) the Jeffrey–Bolker axioms guarantee the existence of closely related pairs of probability and utility functions for any preference ordering satisfying the axioms, (b) the utility functions are order-preserving, and (c) the causal counterexamples to V-maximization show that the Jeffrey–Bolker utilities mismeasure the agent's preferences in those situations, it is clear that the agent's preferences violate one or more of the axioms of the Jeffrey–Bolker theory. It strikes me that among the axioms the only plausible candidate for a violation that accounts for the causal counterexamples is the Impartiality condition.[13] Getting clear about exactly how Impartiality is violated in these problems should be a useful step in the project.

2. Another axiom important to the Jeffrey–Bolker theory is their Averaging condition,[14] which is like an important axiom of Fishburn's (see Axiom 9). In both theories the axioms seem quite reasonable. Can their presence in the two theories be exploited?

These points seem worth noticing if it is the Jeffrey–Bolker theory that one seeks to adapt to get the kind of foundation under discussion. The other obvious place to look for ideas is the Luce and Krantz theory – in spite of their commitment to the peculiar disjunctive conditional preferences, that theory (or parts of it) may well be useful to the project. Of course, the idea of looking to Jeffrey–Bolker has been strongly suggested to me by the points Skyrms and Jeffrey have made about applying it to causal decision theory. As I mentioned early in the paper (see note 1), they have both sketched foundations for causal decision theory. In both cases they apply Jeffrey–Bolker theory in combination with an assumed prior specification of appropriate K-partitions. I would like to see a foundation as elegant as Jeffrey–Bolker that preserves the solution to the problem of selecting K-partitions I have presented above. Until one is given, I think the foundation presented here (based on Fishburn's theory, let me emphasize) is the best available.

Department of Philosophy
Ohio State University

NOTES

[1] It is the second of the virtues that I want to emphasize for this foundation: I am aware of two other sketches of foundations for causal decision theory which perhaps share the first virtue. Richard Jeffrey presents one in the concluding section of his (1981), and Brian Skyrms gives another in the concluding section of his (1982). Both suggestions depend on a prior specification of appropriate K-propositions (dependency hypotheses), and so they lack the second virtue described more fully below.

[2] The Twin Prisoners' Dilemma is a standard Prisoner's Dilemma, with the additional assumption that each prisoner believes that the other prisoner makes choices very much as he does. So each prisoner believes that there is a strong statistical correlation between his choice and the other's choice. For prisoner A, Pr(B rats / I rat) is high, and Pr(B cooperates / I cooperate) is high. Likewise for B. Each prisoner still believes, however, that his choice and action do not causally influence the other's choice and action. See Jeffrey (1983), Example 11 in Chapter 1.

[3] Suppose I value the pleasure I would derive from smoking this cigarette (act A_s) at 2 utiles, while I attach a large negative utility to contracting cancer, -1000 utiles. My utility function might say $V(S) = 2$, $V(-S) = 0$, $V(X) = -1000$, $V(-X) = 0$, where S is "I enjoy this cigarette," and X is "I contract cancer." Suppose I have these degrees of belief, where A_r is the act of refraining from smoking the cigarette:

$$\begin{array}{ll} P(S \,\&\, X/A_s) = 0.7 & P(S \,\&\, X/A_r) = 0.01 \\ P(S \,\&\, -X/A_s) = 0.25 & P(S \,\&\, -X/A_r) = 0.04 \\ P(-S \,\&\, X/A_s) = 0.04 & P(-S \,\&\, X/A_r) = 0.25 \\ P(-S \,\&\, -X/A_s) = 0.01 & P(-S \,\&\, -X/A_r) = 0.7 \end{array}$$

(J) tells us that

$$\begin{aligned} V(A_s) &= \sum_i P(C_i/A_s)\, V(C_i \,\&\, A_s) \\ &= 0.7(-998) + 0.25(2) + 0.04(-1000) + 0.01(0) \approx -738 \quad \text{and} \\ V(A_r) &= 0.01(-998) + 0.04(2) + 0.25(-1000) + 0.7(0) \approx -260 \end{aligned}$$

[4] See Stalnaker (1972) for the suggestion these theories build on.

[5] It may be that the agent is unsure about the causal structure of his decision problem:

he may have partial belief in a number of hypotheses about it. It is important to note that causal expected utility theory adequately handles these more complicated and more complicated and more realistic decision problems. The idea is simple: build his various hypotheses about the causal story into his appropriate K-partition. The hypotheses describe states of affairs beyond his influence which are relevant to the outcomes of his action. Each hypothesis will suggest a partition of factors which, according to the hypothesis, are appropriate K's; the expanded partition whose members are conjunctions of these factors with the hypotheses will be an appropriate K-partition. For details on this, see Skyrms (1979, pp. 136–138). The idea goes back to Savage (1954).

[6] I regard the sophisticated V-maximization theories as very interesting, but not because they are theories as good as causal decision theory. Only unreasonably strong assumptions about the agent's self-knowledge and about his fallibility in executing his chosen actions will save these V-maximization theories from *approximating* the correct answers that causal decision theory gives, or from getting the correct answers *almost* all the time. See Jeffrey (1981, 1983) and Eells (1982) for the variations.

[7] Eells (1982).

[8] In the following discussion I will sometimes use the phrase, "the agent has a preference for x" to mean simply "x appears in the agent's preference ranking," rather than "x ranks higher than some y in the agent's preference system.

[9] So like all such theories we fail here to provide a "fundamental measurement" of preference, since the mixing coefficients build some probability structure into the preferences from the beginning. I will discuss this point and the possibility of giving a mixture-free foundation below.

[10] A more extensive treatment of the notion of conditional preference appears in Armendt (1986b). An important point made there is the following: A preference P,Q is the agent's preference for P, under the supposition that Q holds. And the supposition that Q is one in which the agent's beliefs about the causal structure of the world are minimally altered (unless explicitly cancelled by Q). The supposition that Q is not a supposition that Q occurs no matter what the agent does, for example (though the agent might make such a supposition; the point here is that it is a *different* supposition from simply supposing that Q). I take this tendency to preserve the agent's beliefs about causal structures – the ways states, acts, and consequences produce each other in the actual world – as entirely appropriate in a theory of conditional preference tied to rational *action*, where the agent is interested in producing (in the actual world) desirable actions and consequences.

[11] In the precise statement of this rule given below, $P(B/B \vee C)$ is written $P_{B \vee C}(B)$. The representation theorem actually finds probability measures P_Q on subsets of $\mathscr{E}$ for every proposition Q in $\mathscr{E}'$. It establishes that these probabilities behave as conditional probabilities; see clause (iii) of the theorem. The probability measure P in the text is the measure P_T.

[12] The Luce and Krantz (1971) theory also has a purely qualitative preference ordering, but as I mentioned before, it incorporates disjunctive conditional preferences that I find very problematic.

[13] The Impartiality condition states that if A, B, and C are pairwise disjoint, and $A \sim B \sim C$, and $(A \vee C) \sim (B \vee C)$, then for all D disjoint from A and from B, $(A \vee D) \sim (B \vee D)$. See Jeffrey (1983), p. 147.

[14] Averaging says that if A and B are disjoint, and $A \geqslant B$, then $A \geqslant A \vee B \geqslant B$. See Jeffrey (1983), p. 146.

REFERENCES

Armendt, B. (1983) *Rational Decision Theory: The Foundations of Causal Decision Theory*. Ph.D. dissertation for the Department of Philosophy, University of Illinois at Chicago.

Armendt, B. (1986a) 'A foundation for causal decision theory', *Topoi*, **5**, 3–19.

Armendt, B. (1986b) 'Conditional preference and rational choice.' Presented at the 1986 meetings of the Pacific Division of the American Philosophical Association.

Balch, M. and Fishburn, P. (1974) 'Subjective expected utility for conditional primitives', in *Essays on Economic Behavior under Uncertainty*, Balch, McFadden, and Wu (eds.), North-Holland.

Bolker, E. (1966) 'Functions resembling quotients of measures', *Transactions of American Mathematical Society* **124**, 292–312.

Bolker, E. (1967) 'A simultaneous axiomatization of utility and subjective probability', *Philosophy of Science* **34**, 333–340.

Eells, E. (1982) *Rational Decision and Causality*, Cambridge University Press.

Fishburn, P. (1973) 'A mixture-set axiomatization of conditional subjective expected utility', *Econometrica* **41**, 1–25.

Gibbard, A. and Harper, W. (1976) 'Counterfactuals and two kinds of expected utility', in *Ifs*, Harper, Stalnaker, and Pearce (eds.), Reidel.

Herstein, I. and Milnor, J. (1953) 'An axiomatic approach to measurable utility', *Econometrica* **21**, 291–297.

Jeffrey, R. (1965) *The Logic of Decision*, McGraw Hill.

Jeffrey, R. (1981) 'The logic of decision defended', *Synthese* **48**, 473–492.

Jeffrey, R. (1983) *The Logic of Decision*, 2nd edition, University of Chicago Press.

Krantz, D., Luce, R. D., Suppes, P., and Tversky, A. (1971) *Foundations of Measurement*, Vol. I, Academic Press.

Lewis, D. (1981) 'Causal decision theory', *Australasian Journal of Philosophy* **59**, 5–30.

Luce, R. D. and Krantz, D. (1971) 'Conditional expected utility', *Econometrica* **39**, 253–271.

Savage, L. J. (1954) *The Foundations of Statistics*, Wiley.

Skyrms, B. (1979) *Causal Necessity*, Yale University Press.

Skyrms, B. (1982) 'Causal decision theory', *Journal of Philosophy* **79**, 695–711.

Skyrms, B. (1984) *Pragmatics and Empiricism*, Yale University Press.

Sobel, J. H. (1978) *Probability, Chance, and Choice: A Theory of Rational Agency*, unpublished.

Stalnaker, R. (1972) 'Letter to David Lewis', in *Ifs*, Harper, Stalnaker, and Pearce (eds.), Reidel.

von Neumann, J. and Morgenstern, O. (1947) *Theory of Games and Economic Behavior*, 2nd edition, Princeton University Press.

WILLIAM L. HARPER

CAUSAL DECISION THEORY AND GAME THEORY: A CLASSIC ARGUMENT FOR EQUILIBRIUM SOLUTIONS, A DEFENSE OF WEAK EQUILIBRIA, AND A NEW PROBLEM FOR THE NORMAL FORM REPRESENTATION

The standard cases for motivating causal decision theory are ones like Newcomb's problem and the Prisoners' Dilemma where an act can count as evidence for a desired state even when you know you can't influence the objective chance that the state obtains by choosing the act. Strategic reasoning in normal form games provides a rich source of examples where this combination of causal independence and epistemic dependence is met. In Section 1 I suggest that a causal independence assumption is part of the idealization built into the normal form. The Death in Damascus story is offered as an example of a game where this independence assumption is clearly met and where standard game theoretic reasoning applies. This game is used to motivate a conception of ratifiable choice as explicated in causal decision theory.

In Section 2 it is argued that this conception of ratifiable choice can be combined with best response reasoning to explicate the classic *a priori* argument for Nash equilibria. According to this explication the argument can apply even if the equilibria are weak. One application is an answer to an argument by McClennen (1978) which purports to show that game theory is inconsistent with its decision theoretic foundation. Another is a new criticism of Harsanyi's (1977) suggestion that maximin strategies ought to be played when no equilibrium offers higher payoffs. Yet another, is an answer to a recent criticism by Margaret Gilbert (1984) of David Gauthier's (1975) suggestion that rational agents in a game can be expected to reach a unique best point if it is also the unique Nash equilibrium.

In Section 3 this explication of the classic *a priori* argument is developed into a concept of best response admissibility. A theorem is proved showing that *only* Nash strategies are best response admissible in two-person non-cooperative games. A second theorem shows that *all* Nash strategies are best response admissible in any two-person non-

William L. Harper and Brian Skyrms (eds.), Causation in Decision, Belief Change, and Statistics, II, 25–48.

cooperative game where the independence assumption built into the normal form is satisfied. Some suggestions are made about how the causal decision theoretic framework can be further exploited to extend the solution concept for some of these normal form games.

In Section 4 the causal decision theoretic framework is used to illuminate some of the recent criticisms of the normal form that have grown out of Selten's (1965) discussion of the problem of non-credible threats. I think cases where these criticisms apply involve violation of the causal independence assumption built into the normal form. The plot from Puccini's opera *Gianni Schicchi* motivates a striking example where this problem arises. In this extensive form game only one of the large class of Nash equilibria is best response admissible. A variation on the story provides an alternative extensive form game *with the same normal form* where the causal independence assumption is met. These two extensive form games have very different analyses in the causal decision theoretic framework.

1. UNSTABLE CHOICE AND RATIFIABILITY IN NORMAL FORM GAMES

It is illuminating to formulate the Death in Damascus problem (Gibbard and Harper, 1978, p. 156) as a zero-sum game. You opponent is Death. Your pure options are to stay in Damascus $A1$ or to go to Aleppo $A2$. Her options are to seek you in Damascus $B1$ or to seek you in Aleppo $B2$. If you are in the place where she seeks you, you die. If not, you get a reprieve. Let -100 be assigned as your utility for meeting Death and 0 as your utility for a reprieve.

Game 1.1

$$\begin{array}{c} \\ A1 \\ A2 \end{array}\begin{array}{c} \begin{array}{cc} B1 & \quad B2 \end{array} \\ \begin{bmatrix} -100 & 0 \\ 0 & -100 \end{bmatrix} \end{array}$$

You believe Death is very good at predicting your choice. You assign the following epistemic conditional probabilities.

$$\mathrm{P}(B1|A1) \approx 1 \approx \mathrm{P}(B2|A2)$$

You also believe that Death doesn't cheat. Her choices are made independently of yours, perhaps already made on the basis of her accurate reading of your character and circumstances. The independence assumptions built into normal form games are met.

One of the main assumptions of classical game theory is that strategic reasoning in any non-cooperative game that meets the idealizations built into the extensive form representation can be adequately represented in the normal form (e.g. Luce and Raiffa, pp. 52–53). This assumption has recently come under quite a bit of criticism (e.g. Selten, 1965, 1975; Krepps and Wilson, 1982). In Section 4 I shall say something about how causal decision theory can illuminate some of the problems with the normal form. For the present, I want to show how causal decision theory can illuminate games (such as our Death in Damascus example) where the normal form is appropriate.

The basic idea built into the normal form is that rational players with complete knowledge of the game structure and of the utilities of all the players ought to be able to choose their strategies (a strategy is a complete plan for how you would play out the game under any of the contingencies it would allow for) ahead of time – without any appeal to *a posteriori* knowledge of the strategy any other player actually ends up playing out. This independence assumption requires that each player be able to regard the strategies that will be executed by the other players as causally independent of his or her own choice of strategy.

Causal decision theory provides a framework within which this independence requirement can be explicitly represented. On the Gibbard–Harper formulation the appropriate probabilities to use in evaluating expected utility are probabilities of subjunctive conditionals

$$A \;\Box\!\!\rightarrow\; Bi$$

where the antecedent is the alternative you are evaluating and the *Bi*'s are the relevant outcome determining states. In a two-person game the *Bi*'s are the pure strategies the other player might end up playing out. According to this formulation of causal decision theory your unconditional expected utility for alternative strategy *A* is

$$\underset{i}{\text{SUM}}\; \text{P}(A \;\Box\!\!\rightarrow\; Bi)\cdot \text{u}(A,Bi),$$

where each

$$\text{P}(A \;\Box\!\!\rightarrow\; Bi)$$

is your present unconditional epistemic probability for the subjunctive conditional

> If I were to play strategy A the other player would end up playing out pure strategy Bi.[1]

If you believe that your opponent's choice of strategy and the pure strategy he will end up actually playing out are causally independent of which strategy you play then you will assign

$$P(A \mathrel{\Box\!\!\rightarrow} Bi) = P(Bi)$$

The independence assumption of normal form games is represented by the requirement that this belief in causal independence extends to conditional expectations so that

$$\text{(assumption I)} \quad P(A' \mathrel{\Box\!\!\rightarrow} Bi \mid A) = P(Bi \mid A)$$

for any alternative strategies A' and A. In our Death in Damascus example this is to require that on the assumption that I execute $A1$ (staying in Damascus) I assume that I would have escaped if I had chosen $A2$ (going to Aleppo) instead.

Richard Jeffrey (1983, p. 18) has proposed ratifiability as a requirement on rational decisions. His basic idea is that an act A is *ratifiable* just in case no alternative has a higher expected utility on the assumption that A is chosen than A itself has. Explicating this idea requires some way to represent the evaluation of an alternative A' from the epistemic point of view corresponding to the assumption that A is performed. Causal decision theory provides a natural framework for this. According to the Gibbard–Harper formulation you can evaluate the causal utility of another alternative A' from the epistemic point of view corresponding to the assumption that you will choose strategy A:

$$U_A(A') = \underset{i}{\text{SUM}}\, P(A' \mathrel{\Box\!\!\rightarrow} Bi \mid A) \cdot u(A', Bi)$$

The appropriate evaluation is the conditional expectation on A of the causal utility of alternative A'. This supports the following explication of Jeffrey's intuitive idea:

$$A \textit{ is ratifiable} \text{ iff } U_A(A) \geqslant U_A(A') \text{ all } A'$$

The problem with the pure strategies in Death in Damascus is that they are unratifiable in the sense of this explication of Jeffrey's idea.[2]

Our representation of the independence assumption built into the normal form (assumption I) allows the subjunctive conditionals to drop out so that the relevant conditional expectation reduces to the following very simple formulation

$$U_A(A') = \underset{i}{\mathrm{SUM}}\ P(Bi|A) \cdot u(A',Bi)$$

Under this independence assumption both of your pure strategies are unstable. Suppose you decide to stay in Damascus. This gives you evidence that Death will seek you there. Therefore, deciding to stay puts you in an evidential position where going to Aleppo is evaluated as the better option. But, the same problem comes up in reverse if you decide to go to Aleppo instead.

I assume that Death cannot predict the outcome of the chance device that an agent uses to execute a mixed strategy even though she can predict which mixed strategy gets chosen. According to standard zero-sum game theory the mixed strategy $\langle 1/2\ A1, 1/2\ A2 \rangle$ is the best thing you can do in this terrible situation. It guarantees you an expected utility of -50 whatever strategy pure or mixed Death might choose. Any departure from 1/2 will get you into trouble. If you play a strategy $(x\ A1, (1-x)A2)$ where $x > 1/2 (x < 1/2)$ you can assume that Death will play $B1(B2)$. Under this assumption your expected utility for the mixture $x\ A1$ is worse than your expected utility for the $1/2\ A1$ mixture recommended by game theory. The unratifiability of the non-equilibrium strategies in either version of Death in Damascus is just an example of standard game-theoretic strategic reasoning.

I submit that Ratifiability, in this sense of stability, is a necessary condition on non-pathological applications of causal decision theory. The basic recommendation is:

> Choose from among your ratifiable alternatives one which maximizes unconditional causal utility.[3]

This recommendation for rational choice has interesting implications for game theory.

2. BEST RESPONSE REASONING AND WEAK EQUILIBRIA

If when you evaluate your conditional epistemic probability over the other player's strategies on the assumption that you choose A you assume that the other player chooses some best response to A, you are using best response reasoning to guide your expectations about what that player will do. Almost any introductory lecture on game theoretic equilibria will appeal to such reasoning to illustrate the problem of zero-sum games without pure strategy equilibria. This kind of reasoning is also appealed to in von Neumann and Morgenstern's indirect argument (1953, p. 198) for equilibrium strategies and in Luce and Raiffa's (1957, p. 63) *a priori* argument for the requirement that solutions be equilibria.

To ascribe to such best response reasoning is to make an assumption about the expectations that rational players can entertain about each other's strategies. We can represent your commitment to this assumption as a condition on your prior epistemic probability over the other players strategies.

P *is a best response prior* iff $P(B|A) = 0$ unless B is some best response to A

You accept the best response reasoning assumption by assigning for every strategy A of yours $P(B|A) = 0$ for each of this strategies B that fails to be a best response to A.

Some motivation for such best response reasoning is provided by the back-tracking character of epistemic conditional belief. Suppose you assume there is a unique rational choice in your situation and that you will end up committing yourself to it, but you haven't yet figured out what that choice will be. When you consider the hypothetical news provided by your assumption that you choose strategy A you hypothetically assume that A is the unique rational act. You keep fixed your assumption that what you will end up committing yourself to will be the rational choice, and assume hypothetically that reasoning legislated choosing A. In game theory you assume your opponent to be ideally rational; therefore, you assume she or he will have been able to reconstruct the reasoning that leads you to choose A and will have predicted your choice. Thus, you assume that she or he will choose some best response to A when you assume that you will choose A.

Von Neumann and Morgenstern (p. 148) liken such best response reasoning to indirect proof in mathematics. One way to rule out a hypothesis is to assume it and then derive a contradiction. This is the familiar practice of proof by reductio. Such reasoning is no less legitimate a tool for evaluating what rationality principles commit you to in a choice situation than for evaluating the commitments of a mathematical or physical theory. I see no reason why it should be incoherent or circular to use the hypothetical assumption that *A* is the unique rational choice as part of the reasoning that leads you to rule out strategy *A* as a candidate for this solution to your game.

The real problem with best response reasoning is that it, apparently, is just not strong enough to do the job game theorists want done. It seems to fail to support commitment to weak equilibria – those where the best response fails to be unique. Compare the following two zero-sum games:

Game 2.1

$$\begin{array}{c} \\ A1 \\ A2 \end{array}\!\!\begin{array}{c} \begin{array}{cc} B1 & B2 \end{array} \\ \begin{bmatrix} 1 & 2 \\ 3 & 4 \end{bmatrix} \end{array}$$

Here best response reasoning would successfully mandate the equilibrium (*A*2, *B*1). This equilibrium is strong.[4] Not only is each strategy a best response to the other, it is the unique best response. If you depart from your equilibrium strategy you can expect to end up worse off. The same goes for the other player. Contrast this with the following game:

Game 2.2

$$\begin{array}{c} \\ A1 \\ A2 \end{array}\!\!\begin{array}{c} \begin{array}{ccc} B1 & B2 & B3 \end{array} \\ \begin{bmatrix} 1 & 4 & 1 \\ 1 & 0 & 2 \end{bmatrix} \end{array}$$

Look at it from Row chooser's point of view. Assume column chooser plays the unique maximin strategy *B*1. What is your best response? Apparently it doesn't matter what you do. Either option offers the same expected utility, 1. The Nash equilibria for this game are all the pairs $(B1,(xA,(1-x)A2))$ where $x \geqslant 1/4$. But these equilibria are weak. You (Row chooser) can depart to a non-equilibrium strategy without expecting to lose anything.

McClennen (1978, p. 351) offered Game 2.2 as an example to argue that zero-sum game theory is inconsistent with its decision theoretic foundations. He proposes principles which must be satisfied by any game theory picking out an admissible solution set *T* from the set *S* of all strategies available to you in this game, if that theory is to be consistent with expected utility theory. Suppose for each strategy x, $U(x)$ is your expected utility for x under the assumption that you expect your opponent not to be unreasonable according to the lights of the game theory in question. One principle. (McClennen's C7) tells us that if the expected utility of x is maximal in the whole strategy set then it ought to be included in the solution set.

$$\text{Rule (1)} \quad \text{If } U(x) \geqslant U(x') \text{ all } x' \in S \text{ then } x \in T$$

Another (McClennen's C12) tells us that if the expected utility of x is as good as that of some x' in the solution set then x also ought to be in the solution set.

$$\text{Rule (2)} \quad \text{If } U(x) \geqslant U(x') \text{ some } x' \in T \text{ then } x \in T$$

In Game 2.2 the pure strategy $A2$ is not in your solution set. Under the assumptions of the minimax-theory you (Row chooser) can expect column chooser to select $B1$. Therefore $U(A2) = 1$, which is as high as the expected utility of any alternative in your whole strategy set. Thus, Rule 1 is violated. Rule 2 is violated as well, since $U(A2) = U(A1)$ and $A1$ is in your solution set.

The basic recommendation of causal decision theory provides an expected utility framework that lets the best response reasoning assumption support commitment to weak equilibria. It thus provides an answer to McClennen's challenge. Under the best response reasoning assumption in Game 2.2 your non-Nash alternative $A2$ is not ratifiable.

$$U_{A2}(A2) = 0 < U_{A2}(A1) = 4,$$

since $P(B2 \mid A2) = 1$. You already expect column chooser to choose $B1$. Your hypothetical assumption that you choose $A1$ gives you no hypothetical grounds to change this, since $B1$ is a best response to $A1$. Therefore $P(B1 \mid A1) = 1 = P(B1)$ and $A1$ is ratifiable.

$$U_{A1}(A1) = 1 = U_{A1}(A) \text{ all } A$$

The same holds for any strategy in the solution set. $B1$ is a best response to $(x\ A1,\ 1-x\ A2)$ so long as $x \geqslant 1/4$. Therefore, under the assumptions McClennen makes the ratifiable strategies are exactly the set recommended as solutions in game theory. The basic recommendation of causal decision theory is to choose from among the ratifiable options one with maximal unconditional causal utility. In the present case this leads to no further reduction because the unconditional causal utility of each of these options is the same 1.0.

This shows that neither of McClennen's rules is violated if strategic reasoning goes by the best response assumption and expected utility goes by the basic recommendation to choose by causal utility from among your ratifiable options. Best response strategic reasoning makes causal decision theory legislate as admissible exactly the strategies allowed as admissible by game theory.

Another application of this ratifiability defense of weak equilibria is a new criticism of John Harsanyi's (1977, p. 125) suggestion that your maximin strategy should be played when none of your equilibrium strategies offers a higher expected utility than it does. Harsanyi (1977, p. 125) offered the following game to motivate his suggestion:

Game 2.3

$$\begin{array}{c} \\ A1 \\ A2 \end{array}\!\!\begin{array}{cc} B1 & B2 \\ \left[\begin{array}{cc} (60,24) & (0,44) \\ (20,84) & (40,4) \end{array}\right] \end{array}$$

In this game the unique equilibrium strategies are $A^* = (4/5\ A1, 1/5\ A2)$ and $B^* = (1/2\ B1, 1/2\ B2)$. The expected payoffs for the pair (A^*, B^*) are (30,36). This equilibrium is weak in that the payoff to player 2 (column chooser) if player 1 (row chooser) does A^* is 36 whatever strategy player 2 chooses. Similarly, if player 2 chooses B^* player 1 will expect 30 whatever strategy player 1 chooses. The maximin strategy for player 1 is $A^- = (1/4\ A1, 3/4\ A2)$ while that for column chooser is $B^- = (2/5\ B1, 3/5\ B2)$. They lead to the same payoff pair (30,36) as do the equilibrium strategies, but Harsanyi argues that the maximin strategies should be selected:

> Yet player 1 has no reason to expect that player 2 will in fact stick to the equilibrium strategy B^*, because player 2 will not suffer any kind of penalty if he shifts to another strategy while player 1 himself uses strategy A^*. Accordingly Postulate A [Harsanyi's Proposal] suggests that in games of this type the players should use maximin strategies instead of trying to reach an equilibrium point. (p. 125)

This is exactly the sort of reasoning about weak equilibria that led to McClennen's argument that game theory is inconsistent with its decision theoretic foundation. The same answer applies here. Player 1 can expect player 2 to stick to B^* because any departure from B^* is unratifiable by best response reasoning. Suppose player 2 entertains his maximin strategy B. Player 1's best response to that is the pure strategy $A2$ which gives the payoffs (32,36), but player 2's best response to $A2$ is not B but $B1$,

$$U_B(B) = 36 < U_B(B1) = 84.$$

The maximin strategy is not ratifiable under best response reasoning.

One might argue that the best response reasoning assumption should not apply, but even on Harsanyi's own proposal the game has a unique solution. It would seem that rational players ought to be able to figure out what it is. Thus, it seems to me that this game is a reductio of Harsanyi's suggestion rather than positive motivation for it.[5]

Margaret Gilbert (1984) is even more ready to opt for maximin strategies over equilibrium strategies than Harsanyi's suggestion. She uses the following game to argue against David Gauthier's (1975, p. 200) contention that rational agents will do their parts to reach a unique best point which is also the unique equilibrium.

Game 2.4

$$\begin{array}{c} \\ A1 \\ A2 \\ A3 \end{array} \begin{array}{c} \begin{array}{lll} \beta_1 & \quad\quad\quad \beta_2 & \quad\quad\quad \beta_3 \end{array} \\ \left[\begin{array}{lll} (5,5) & (-50,-50) & (2,4) \\ (-50,-50) & (2,4) & (4,3) \\ (4,4) & (3,3) & (3,3) \end{array}\right] \end{array}$$

Here the unique equilibrium is (A_1,β_1) and it is also a unique best point for each player. The maximin strategy A_3 ensures that the worst outcome (-50) is avoided. According to Gilbert (p. 10) this seems to provide a counterweight to any argumentation for A_1. I think that best response reasoning applies in this game so that the ratifiability argument for strategy A_1 overrides the counterweight. Strategy A_3 could not be ratified by any best response prior; therefore, it is not admissible as a candidate for the unique solution to this game. In the next section I shall use best response admissibility to begin construction of a solution concept. This construction does yield a unique solution to Gilbert's game and that solution is (A_1,β_1).[6]

3. NASH EQUILIBRIA AND THE CONCEPT OF A SOLUTION IN NORMAL FORM GAMES

It is possible to have a game and a best response prior where no strategy whatever is ratifiable.

Game 3.1

$$\begin{array}{c} \quad B1 \quad B2 \\ \begin{matrix} A1 \\ A2 \end{matrix} \begin{bmatrix} 2 & 1 \\ 1 & 2 \end{bmatrix} \end{array}$$

The solution to this game is the pair ((1/2*A*1,1/2*A*2), (1/2*B*1,1/2*B*2)). It is easy to show that no non-solution strategy is ratifiable by any best response prior. Now suppose that $P(B1|(1/2A1,1/2A1)) \neq 1/2$. Let it equal 1.0. This is still a best response prior. But, on it your Nash strategy (1/2*A*1,1/2*A*2) will fail to be ratifiable. This will leave you with no ratifiable strategy whatever.[7]

Let us introduce the idea of a best response admissible strategy.

> *x is best response admissible* iff
> some best response prior would ratify it.

A strategy which fails to be best response admissible could not be ratified by any best response prior; therefore, it could not be rationally adopted by an agent who used best response reasoning.

For two-person games a strategy pair (*x*,*y*) are a Nash equilibrium just in cash each is best against the other (Nash, 1950, p. 287). Let us say that *x* is a Nash strategy just in case there is a *y* such that (*x*,*y*) is a Nash equilibrium. We are now in position to articulate a couple of theorems for non-cooperative games.

THEOREM I *In two-person games only Nash strategies are best response admissible.*

Proof. Assume P is some best response prior which ratifies *x*. We have for all alternatives *x*′ to *x*.

1. $$\mathrm{SUM}_i\ \mathrm{P}(x \mathrel{\Box\!\!\rightarrow} Bi|x) \cdot \overset{a}{\mathrm{U}}(x,Bi) \geqslant \mathrm{SUM}_i\ \mathrm{P}(x' \mathrel{\Box\!\!\rightarrow} Bi|x) \cdot \overset{a}{\mathrm{U}}(x',Bi),$$

where $\overset{a}{\mathrm{U}}(x,Bi)$ is the expected utility to player *a* of the result of playing strategy *x* against player *b*'s pure strategy *Bi*.

The following theorem from the logic of subjunctive conditionals,[8]

2. $$A \supset [(A \mathrel{\Box\!\!\rightarrow} B) \equiv B],$$

ensures that

3. $\quad P(x \;\Box\!\!\rightarrow Bi|x) = P(Bi|x),$

even if $P(x' \;\Box\!\!\rightarrow Bi|x) \neq P(Bi|x)$ for some alternative x' so that causal independence fails.

Let y^* be a mixed strategy for player b such that for each i,

4. $\quad yi^* = P(Bi|x)$

We know x is best against y^* by 1 and 3.

Now show y^* is best against x. Where Y is the set of all strategies of player b let Yx be the set of best responses to x, i.e.

5. $$y \in Yx \text{ iff } \mathrm{SUM}_i\, yi \cdot \overset{b}{\mathrm{U}}(x,Bi) \geqslant \mathrm{SUM}_i\, y'i \cdot \overset{b}{\mathrm{U}}(x,Bi) \text{ all } y' \in Y$$

Since P is a best response prior, we have

6. $$\int_{y \in Yx} P(y|x)\,\mathrm{d}p = 1$$

so that

7. $$P(Bi|x) = \int_{y \in Yx} P(y|x) \cdot yi\ \mathrm{d}p$$

for any appropriate integral over Yx.

By 4 and 7 we have,

8. $$\mathrm{SUM}_i\, yi^* \cdot \overset{b}{\mathrm{U}}(x,Bi) = \mathrm{SUM}_i \int_{y \in Yx} P(y|x) \cdot yi\ \mathrm{d}p \cdot \overset{b}{\mathrm{U}}(x,Bi)$$
$$= \int_{y \in Yx} P(y|x) \cdot (\mathrm{SUM}_i\, yi \cdot \overset{b}{\mathrm{U}}(x,Bi))\mathrm{d}p$$

so that y^* is a mixture of best responses to x. Any mixture of best responses is a best response, so y^* is best against x as required.[9]

This theorem shows that the answer to McClennen provided by best response reasoning together with ratifiability formulated in causal decision theory also provides a general argument recommending that solutions to two-person non-cooperative games be restricted to Nash equilibria.[10]

THEOREM II *In two-person games where causal independence (Assumption I) holds* all *Nash strategies are best response admissible.*

Proof. Let (x,y) be a Nash equilibrium. Choose a best response prior P such that,

1. $P(y|x) = 1.$

By Assumption I, we have for all alternatives x' to x.

2. $P(x' \,\square\!\!\rightarrow Bi|x) = P(Bi|x),$

so that

3. $$\overset{a}{U}_x(x') = \text{SUM}_i\ P(Bi|x) \cdot \overset{a}{U}(x',Bi)$$
$$= \text{SUM}_i\ yi \cdot \overset{a}{U}(x,Bi)$$
$$= \overset{a^i}{U}(x',y) \leq \overset{a}{U}(x,y)$$

So, P ratifies x, as required.

This proof appeals to Assumption I. In the next section we shall see an example where this assumption is violated and many Nash strategies fail to be best response admissible.

Nash (1951, p. 288) proved (his famous fixed point proof) that his equilibria exist for any finite game; therefore, it follows that there are best response admissible strategies in any finite two-person game. Where the equilibrium is unique or where the set of equilibria are interchangeable (Nash's strong solution concept, p. 290) it makes sense to say that best response admissibility specifies the solution to a game. For example, in zero-sum normal form games best response admissibility allows exactly the maximin strategies and in these games they are either unique or interchangeable. So, best response reasoning underwrites the standard normal form solution for zero-sum games provided by the minimax theorem. Similarly, best response admissibility specifies the solution to Gilbert's game.

The problem with many games is that they have several non-interchangeable Nash equilibria. Often there are just too many different Nash equilibria to provide an adequate solution concept. This problem is especially serious because it undercuts the best response reasoning that underwrites the restriction of solutions for normal form games to

Nash equilibria in the first place. Unless some way can be found to cut down the set of Nash equilibria for a given game to a subset which satisfies Nash's strong solution concept the best response reasoning used to argue for Nash equilibria may be suspect. Suppose all the rationality considerations there are do not suffice to restrict my strategies to an interchangeable set. This would undercut the original assumption I used in applying the von Neumann–Morgenstern indirect argument. When I hypothetically assumed I would end up choosing strategy A, perhaps all I should have assumed is that A is some strategy in the solution set. But, this will not allow me to hypothetically assume that an opponent who completely understands the game and the demands of rationality will have predicted my choice. My ground for using a best response prior would seem to be undercut and with it my ground for restricting solutions to best response admissible strategies.[11]

More attention has been given to the problem of finding reasons to exclude some Nash equilibria as undesirable than to any other problem in the development of game theory in the years since the heady days in which Luce and Raiffa wrote their classic survey of the field. When we see that the *a priori* argument for Nash equilibria in the first place may be at stake we see that this problem may have even more urgency than many may have expected.

I want to make it plausible that using causal decision theory as the basic framework for individual rationality might be a useful tool to add to the arsenal available to attack this critical problem. The following game corresponding to the matrix of Richter's (1984) problem provides a good examples where there are too many Nash equilibria:

Game 3.2

$$\begin{array}{c} \\ A1 \\ A2 \end{array}\begin{array}{c} \begin{array}{cc} \quad B1 \qquad\qquad & \qquad\qquad B2 \end{array} \\ \begin{bmatrix} (10,10) & (100,100) \\ (100,100) & (-10000,-10000) \end{bmatrix} \end{array}$$

Row player's best response admissible strategies are the two pure strategies and the mixture $x = (0.9244\,A1, 0.0756\,A2)$. Similarly, column player's best response admissible strategies are the pure strategies $B1$, $B2$ and the mixture $y = (0.9244\,B1, 0.0756\,B2)$. We would like to get more of a solution if we can.

The basic recommendation of causal decision theory provides motivation for going further. It tells you to maximize the unconditional causal utility of your ratifiable alternatives. If you can use strategic reasoning to underwrite constraints on your epistemic probability assignment for the other players best response admissible options you may be able to solve the game. Assume you are row chooser. At first glance best response reasoning does nothing to discriminate among *B*1, *B*2 and the mixture *y*. So try assigning equal probability 1/3 to each. When you evaluate the unconditional causal utility of your three admissible options using this probability assignment you find that *A*1 is the unique maximizing choice. Does this show that it is the solution? No it doesn't. What you have here is a reductio of the assumption that *A*1 is the solution. When you apply the same reasoning you used to guide your choice to the other player you predict that he will symmetrically choose *B*1 but *B*1 is not a best response to *A*1. Are there any epistemic probability assignments to *B*1, *B*2 and *y* that will not get you into trouble? Yes there are; they are exactly the ones which make your mixed strategy *x* maximal. When you assume the other player assigns the corresponding epistemic probabilities to *y* our admissible options then you predict that he will choose *y*. This leads to no problem because *x* and *y* are in equilibrium. Therefore, the solution recommended is the intuitively correct pair of mixed strategies.

This suggests that ratifiable choice as explicated in causal decision theory provides a decision theoretic framework within which the solution concept generated by best response reasoning can be extended. I think exciting results will follow from the attempt to see just how far this solution concept can be pushed.[12]

4. PROBLEMS WITH THE NORMAL FORM

In the years since Selten (1965) raised the issue of non-credible threats more and more game theorists have expressed doubts about the adequacy of the normal form representation. The plot of Puccini's opera *Gianni Schicchi* provides a nice example of the sort of problem these game theorists have been worrying about.[13]

For our purposes the relevant part of the story can be rendered as follows. A very wealthy man has just died. His relatives discover that he has willed his fortune to some monks, instead of to them. They call in a fixer: Gianni Schicchi. Schicchi suggests that they let him impersonate the dead man (no one else yet knows that the man has died) and in the presence of appropriate authorities make out a new will giving the fortune to the relatives. Now we have Schicchi disguised as the rich man begin to make out the new will in the presence of the relatives and the appropriate officials, but instead of naming the relatives he names himself (Gianni Schicchi) as beneficiary of the fortune. Now one of the background assumptions is that the state in which this all happens has very strict penalties for anyone found guilty of colluding to interfere with a will. The penalty is loss of a hand together with banishment.

We can represent the game theoretic reasoning that is critical to this story by the following extensive form game. Player *b* represents Gianni Schicchi. His pure options are either *C* to cooperate with the relatives or *D* to defect by naming himself as beneficiary. He gets to move first. Player *a* represents the relatives. Their pure options are either *A* to accuse Schicchi or *B* to back down. These utilities seem plausible; but, a similar analysis will apply for any assignments of utilities that preserve the rankings of outcomes.

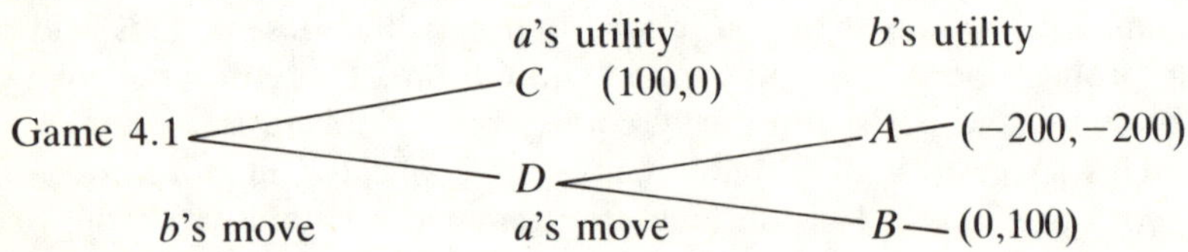

When we look at this extensive form game it becomes apparent that neither *a*'s threat strategy *A* nor *b*'s cooperative strategy *C* are credible. We can use our causal decision theoretic framework to explicitly show that this game has only one equilibrium that can be rationally supported. This is the pure strategy pair

$$(B,D)$$

that corresponds to the actual plot line of the opera.

The normal form analysis for this game uses the idea that there should be no difference between the way *a* should choose between *A* and *B* and the way *a* ought to choose ahead of time between the two conditional strategies

Do *A*, if *b* plays *D*
Do *B*, if *b* plays *D*

The normal form is as follows:

$$\begin{array}{cc} & C \qquad\qquad D \\ \begin{matrix} A \\ B \end{matrix} & \begin{bmatrix} (100,0) & (-200,-200) \\ (100,0) & (0,100) \end{bmatrix} \end{array}$$

Now this normal form game has two pure strategy equilibria (A,C) and (B,D), as well as a set of half mixed strategy equilibria (x,C) where x is a mixture for player a which assigns at least 1/3 probability to strategy A. Our causal theoretic analysis of the extensive form of this game makes use of relevant causal structure that is suppressed in the normal form.

The first step in analyzing the extensive form game in our framework is to determine which strategies are best response admissible. Let us begin with player b.

$$\overset{b}{\mathrm{U}}_C(C) = \mathrm{P}_b(C \,\Box\!\!\rightarrow A\,|\,C)\cdot 0 + \mathrm{P}_b(C \,\Box\!\!\rightarrow B\,|\,C)\cdot 0$$
$$= 0$$

while

$$\overset{b}{\mathrm{U}}_C(D) = \mathrm{P}_b(D \,\Box\!\!\rightarrow A\,|\,C)\cdot -200 + \mathrm{P}_b(D \,\Box\!\!\rightarrow B\,|\,C)\cdot 100$$

How do we evaluate

$$\mathrm{P}_b(D \,\Box\!\!\rightarrow A\,|\,C)?$$

That is how ought b to evaluate the subjunctive

If I had played D instead, a would have played A

from the epistemic point of view corresponding to his assumption that he plays strategy C. Well, how ought he to evaluate

$$\mathrm{P}_b(D \,\Box\!\!\rightarrow A)$$

simpliciter? Clearly a's best response to D is B, since 0 beats -200. Moreover, a gets to move after a knows what player b's move is. Therefore, this is not a case of causal independence. Indeed, it seems reasonable to have b assign

$$\mathrm{P}_b(D \,\Box\!\!\rightarrow A) = 0 \text{ and } \mathrm{P}_b(D \,\Box\!\!\rightarrow B) = 1.$$

Furthermore, it seems apparent that this causal dependence built into the game structure is not changed by b's assumption that he will actually execute C, so relative to the evidential point of view corresponding to this assumption b ought to believe that a would have known it if b had chosen D instead. This supports the assignments

$$P_b(D \,\Box\!\!\rightarrow A \,|\, C) = P(D \,\Box\!\!\rightarrow A) = 0$$

$$P_b(D \,\Box\!\!\rightarrow B \,|\, C) = P(D \,\Box\!\!\rightarrow B) = 1$$

Therefore,

$$\overset{b}{U}_C(D) = 100, \text{ while } \overset{b}{U}_C(C) = 0$$

so, C is not ratifiable by any best response prior that respects the causal structure built into the extensive form of the game.

What about strategy D? By reasoning similar to that we have been exploring we have

$$P_b(D \,\Box\!\!\rightarrow A \,|\, D) = P(D \,\Box\!\!\rightarrow A) = 0$$

$$P_b(D \,\Box\!\!\rightarrow B \,|\, D) = P(D \,\Box\!\!\rightarrow B) = 1$$

So, we have

$$\overset{b}{U}_D(D) = 100 > \overset{b}{U}_D(C) = 0$$

and indeed

$$\overset{b}{U}_D(D) \geqslant \overset{b}{U}_D(y)$$

for any mixture $(yC,(1-y)D)$. Thus D is ratifiable. We also have that no mixture $(yC,(1-y)D)$ is ratifiable other than the trivial one $y = 0$ which is equivalent to the pure strategy D, since for any best response prior that respects the causal structure of the game will have

$$\overset{b}{U}_y(D) = 100 > \overset{b}{U}_y(y) \text{ if } y > 0.$$

Thus, it turns out that the pure strategy D is the only best response admissible strategy for b in this game.

What about player a? Well, a's decision is based on choosing from among a's best response admissible strategies one which optimises a's unconditional causal utility after a has adjusted a's prior to rule out all of b's inadmissible strategies. As we have seen, D is the only best response admissible strategy for b, so a can be sure that b will play D. Thus, the choice between pure strategies A and B for a reduces to a choice between -200 and 0, so that only the pure strategy B could be rationally supported. If B is best response admissible, then our framework will yield the equilibrium pair

$$(D,B)$$

as the unique solution to this game.

Is B best response admissible? Yes it is.

$$\mathrm{P}_a(B \;\Box\!\!\rightarrow D\,|\,B) = \mathrm{P}_a(D\,|\,B) = 1 = \mathrm{P}_a(x \;\Box\!\!\rightarrow D\,|\,B)$$

for any mixed strategy $(xA,(1-x)B)$. This is so because b's choice of D is already settled before hand and therefore cannot be changed by a's choice among A, B or any mixture of them. This has the effect that

$$\overset{a}{\mathrm{U}}_B(B) = 0 \geqslant \overset{a}{\mathrm{U}}_B(x)$$

for any strategy $(xA,(1-x)B)$. Our framework does indeed yield (B,D) as the unique solution to Game 4.1.

In Theorem II we showed that in two-person normal form games all the Nash strategies are best response admissible, provided assumption I holds. In the extensive form of Game I we see that not all Nash strategies are best response admissible. The theorem relies on the causal independence assumption built into the normal form. The extensive form of this game contains causally relevant structure which violates this independence assumption.

Now let us consider a variation of the Puccini story. Suppose the relatives consider the possibility of Schicchi's duplicity and make available to themselves a new strategy A which would be implemented by instructing a confederate ahead of time to blow the whistle on Schicchi (and thus on the relatives as well) if Schicchi defects. The new strategy B is to not set up the whistle blower. On this version of the story the relatives move first, but Schicchi doesn't have available *a posterori* knowledge of which move they've made. The extensive form for this version is as follows:

Game 4.2

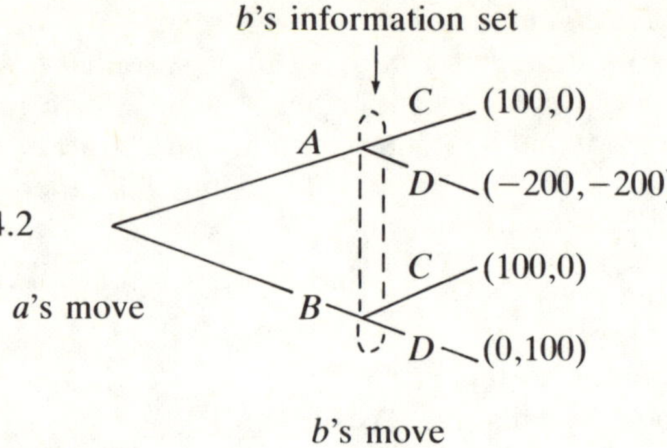

This game has exactly the same normal form as Game 4.1, but here the normal form does not suppress any structure relevant to the causal decision theoretic analysis.

The causal independence assumed in the normal form representation really does hold in this game. Thus, Theorem II applies and the best response admissible strategies are exactly the Nash equilibrium strategies. So, the best response admissible strategies for *b* are the two pure strategies *C* and *D*, and the best response admissible strategies for *a* are the two pure strategies *A* and *B* and all mixtures $(xA,(1-x)B)$ where $x \geqslant 1/3$.

Nash's strong solution concept is not satisfied by this initial best response admissibility analysis, so we need to look at priors. Each player is to optimize that unconditional causal utility among their admissible strategies. Some possible pairs of priors would have

$$P_a(C) = 1 \text{ and } P_b(A) > 1/3$$

For such agents the equilibrium point

$$(A,C)$$

might be appropriate. This equilibrium is destroyed by admitting any uncertainty about *b*'s choice into *a*'s prior (e.g. by a trembling hand assumption (Selten, 1975) or as in Harsanyi's (1975) logarithmic tracing procedure). Perhaps, one could argue that it could not be rational for *a* to be certain that *b* would cooperate in any such game. Nevertheless, I think it should be clear that rather different and in part rather more delicate, considerations are required to mandate (B,D) as the unique solution to Game 4.2 than were required to mandate it for Game 4.1.

This indicates, I think, that the structure suppressed in the normal form can be relevant to the analysis of a game.

Recently, Kohlberg and Mertens (1986) have defended the normal form against the rising tide of opposition generated by the growing appreciation of Selten type difficulties.[14] They argue that solutions ought to be invariant over certain transformations of extensive form games. Invariance over these transformations will require the same solution to all games having the same normal form. Best response admissibility is not invariant over all these transformations; however, best response admissibility is not proposed as a full fledged solution concept. It is a first step to be followed up by using constraints on priors to select from among best response admissible options. It is an open question whether any natural procedure of this sort would lead to a solution concept that would be invariant over all the transformations Kohlberg and Mertens want.

Even if this were to turn out to be so it would not show that the normal form is the best framework for analyzing any given game. As the first version of the *Gianni Schicchi* game shows, the information provided in the extensive form can often let best response admissibility rule out many undesirable equilibria. Just as with the backward inductive reasoning used in applying Krep's and Wilson's (1982) sequential equilibrium concept, so it appears that the extensive form is the natural framework for carrying out best response admissibility analysis. Considerations of this sort could also give some grounds for rejecting one or more of the invariance requirements Kohlberg and Mertens propose.

Talbot College
University of Western Ontario

NOTES

[1] The appropriate evaluation of your epistemic probability of the subjunctive conditional ought to go by your epistemic expectation over relevant hypotheses Hj about the objective conditional chance of Bi on A so that

$$\mathrm{P}(A \;\Box\!\!\rightarrow Bi) = \mathrm{SUM}_j\ \mathrm{P}(Hj) \cdot Cj(Bi|A)$$

where $Cj(Bi|A)$ is the objective conditional chance of Bi given A according to chance hypothesis Hj. One requirement on appropriate chance hypotheses is that they be causally independent of whether you execute option A or any one of its alternatives. In Harper (1986a) I argued that in the context of the independence assumptions of normal form games the mixed strategies of the other player provide a natural set of alternative chance hypotheses so that the subjunctive conditional formulation can be replaced by an appropriate integral representing your expectation over these mixed strategies of the objective chance that *Bi* will end up getting played.

[2] Jeffrey's own explication is different. For a discussion of why this explication is better see Harper (1986b).

[3] See Harper (1986a) for further defense of this recommendation.

[4] This use of "strong" to characterize such equilibria accords with that of Harsanyi (1982, p. 183).

[5] Harsanyi has independently given up this solution concept (personal communication). The important theory he and Selten are developing in their forthcoming book does not allow for such violations of the requirement that solutions be Nash equilibria.

[6] I am grateful to Christina Bicchieri for pointing out to me the value of having an explicit argument for the Nash solution in games like Gilbert's.

[7] I am grateful to Cliff Hooker for first showing me that there are games where this happens even with an "ignorance" prior over best responses.

Brian Skyrms (1986) has been investigating an interesting concept of dynamic deliberation equilibria. Think of an agent calculating expected utility, tentatively inclining toward the better options and changing his probability assignments to his options accordingly, then recalculating expected utilities taking into account those new probabilities he has assigned to his options, etc., until the process settles down on some deliberational fixed point. If that fixed point is a non-trivial mixture over the basic options we have a ratifiable mixed strategy. Skyrms (1986, p. 62) lays down conditions that ensure that such deliberational fixed points always exist. One of these is that updating is to go by Jeffrey rule shifts as you revise your probability assignments to your basic options. If these basic options are your pure strategies and you are row chooser in Game 3.1 and $P(B1|(pA1; (1-p)A2))$ is to be the degree of belief in $B1$ that corresponds to arriving at the assignment $(pA1; (1-p)A2)$ at some stage in dynamical deliberation; then this requires that

$$P(B1|(pA1; (1-p)A2)) = p \cdot P(B|A1) + (1-p) \cdot P(B1|A2).$$

One nice feature of this requirement is that it would rule out the best response prior that led to no ratifiable strategy in Game 3.1. The apparent difficulty with it is that it does not satisfy best response reasoning. For example, it would demand that

$$P(B1|(0.6A1; 0.4A2)) = 0.6,$$

even though $B1$ is the unique best response to the mixture $(0.6A1; 0.4A2)$. More recently Skyrms (1987) has been investigating versions of dynamic deliberation more appropriate to game theory. These investigations promise very exciting results. Indeed, one version would appear to lead rather naturally to an argument restricting solutions to *n*-person games to equilibria.

[8] See Gibbard and Harper 1978, pp. 128–129. See also the introduction and relevant articles in Harper, Stalnaker, Pearce eds. (1980) for more on the logic of subjunctive conditionals. In Gibbard and Harper 1978, we pointed out that $P(A \square\!\!\rightarrow B|A) = P(B|A)$ if $P(A) > 0$. If $P(A) = 0$, but conditional probabilities on A are well defined anyway (as in the extended conditional probability functions defined in Harper 1978 the result still holds, because $P(A \supset [(A \square\!\!\rightarrow B) \equiv B]|A) = 1$ even if $P(A) = 0$.

[9] I am grateful to Haim Gaifman for showing me that the extension to the *n*-person case may be non-trivial and to him and Ken Binmore for helpful suggestions. Brian Skyrms' new work (1987) may offer the best hope (see Note 7).

[10] Recent interesting papers by Pearce (1984) and Bernheim (1984) have shown that common knowledge of best response behaviour is not sufficient for mandating Nash equilibria. This suggests that the classic argument for Nash equilibria needs some explication. For the two-person case this theorem shows that best response admissibility adds enough to provide an explication that works. If an appropriate extension to the *n*-person case can be found then the classic justification for Nash equilibria can be explicated in the full generality intended by Von Neumann and Morgenstern. I am grateful to John McMillin for first bringing the work of Pearce and Bernheim to my attention and to McMillin, Ariel Rubenstein, Christina Bicchieri, and, especially, to Philip Reny for helping me appreciate the relationship of this work to mine.

[11] Wlodzimerz Rabinowicz (circulated manuscript) has attempted to do without the assumption that rationality mandates Nash's strong solution concept. These suggestions lead to quite radical departures from standard solutions, even in two-person zero-sum games. I think this is too much to give up. I prefer to continue to attempt to get a theory that supports the Nash strong solution concept, at least until it is much clearer than it presently is that such attempts are hopeless.

[12] My early attempts to work out details of such a procedure benefited from discussions with Alan Gibbard.

[13] I am grateful to John McCarthy for suggesting that opera plots would be illuminating to investigate and for telling me the *Gianni Schicchi* story. Peter Woodruff offered me an illuminating rendition of some critical arias, and Christopher Brown helped me get the spelling right.

I used this game to illustrate an application of Harsanyi's tracing procedure in comments on Professor Harsanyi at the 1986 Pacific division meeting of the American Philosophical Association.

I am grateful to Bas van Fraassen for a very helpful discussion of this material and especially for convincing me to include the story (and not just the game tree) in the paper.

[14] I am grateful to Philip Reny for helping me get more clear on these relationships to the work of Kohlberg and Mertens. I thank John Harsanyi, Ig Horstman and John McMillin for making versions of the Kohlberg Mertens work available to me. Finally I want to thank John McMillin for his encouragement throughout.

REFERENCES

Bernheim, B. Douglas (1984) 'Rationalizable strategic behavior', *Econometrica* **52**, 1007–1028.

Gauthier, David (1975) 'Coordination', *Dialogue*.

Gibbard, Alan and Harper, William (1978) 'Counterfactuals and two kinds of expected utility', in Hooker *et al.* Vol. I: 125–262.

Gilbert, Margaret (1983) 'Some limitations of rationality', Colloquium paper presented at A.P.A. meeting, Boston 1983.

Harper, William (1978) 'Bayesian learning models with revision of evidence', *Philosophia* 7(2), pp. 357–367.

Harper, William, Stalnaker, Robert, and Pearce, Glen (eds.) (1980) *Ifs*, D. Reidel, Dordrecht.

Harper, William (1986a) 'Mixed strategies and ratifiability in causal decision theory', *Erkenntnis* **24**, 25–36.

Harper, William (1986b) 'Ratifiability and causal decision theory comments on Eells' and Seidenfeld', *P.S.A. 84*. Vol. 2.

Harsanyi, John (1975) 'The tracing procedure: A Bayesian approach to defining a solution for *n*-person games', *International Journal of Game Theory*.

Harsanyi, John (1976) *Essays on Ethics, Social Behavior and Scientific Explanations*, D. Reidel, Dordrecht.

Harsanyi, John (1977) *Rational Behaviour and Bargaining Equilibrium in Games and Social Situations*, Cambridge University Press.

Harsanyi, John (1982) 'Solutions for some bargaining games under the Harsanyi-Selten solution theory; Part I', *Mathematical Social Sciences* **3**, 179–191.

Harsanyi, John and Selten, R. (forthcoming) *A General Theory of Equilibrium Selection in Games*, MIT Press.

Hooker, Clifford A., Leach, James J., and McClennen, Edward F. (eds.) (1978) *Foundations and Applications of Decision Theory*, Western Ontario Series, Number 13. D. Reidel Publishing Co. Dordrecht, Holland.

Jeffrey, R. C. (1983) *The Logic of Decision* (2nd ed.), Chicago and London: University of Chicago Press.

Kohlberg, Elon and Mertens, Jean-François (1986) 'On the strategic stability of equilibria', *Econometrica* **54**, 1003–1037.

Kreps, David M. and Robert Wilson (1982) 'Sequential equilibria', *Econometrica* **50**(4), 863–894.

Luce, Duncan and Raiffa, Howard (1957) *Games and Decisions*, New York: Wiley.

McClennen, Edward F. (1978) 'The minimax theory and expected utility reasoning', in Hooker *et al*. Vol. I: 337–359.

Nash, John F. (1951) 'Non co-operative games', *Annals of Mathematics* **54**, 286–295.

Pearce, David (1984) 'Rationalizable strategic behavior and the problem of perfection', *Econometrica* **52**, 1029–1050.

Richter, Reed (1984) 'Rationality revisited', *Australasian Journal of Philosophy* **62**, 392–400.

Rabinowicz, Wlodzimierz (1985) 'On game theory', Circulated Manuscript.

Selten, R. (1965) 'Spieltheoretische Behandlung eines Oligopolmodels mit Nachfragetragheit', *Zeitschrift für die gesamte Staatswissenschaft* **121**, 301–324.

Selten, R. (1975) 'Re-examination of the perfectness concept for equilibrium points in extensive games', *International Journal of Game Theory* **4**, 22–55.

Skyrms, Brian (1986) 'Deliberational equilibria' *Topoi* **5**, 59–67.

Skyrms, Brian (1987) 'Deliberational dynamics and the foundations of Bayesian game theory', forthcoming in J. Tomberlin (ed.), *Philosophical Perspectives* (Vol. 2): *Epistemology*, Richview Publishing Company.

ERNEST W. ADAMS

CONSISTENCY AND DECISION: VARIATIONS ON RAMSEYAN THEMES

1. INTRODUCTION

This paper will develop ideas of Ramsey's concerning the notion of *consistency* as it applies to systems of decisions and of degrees of belief, which are set forth in the final two sections of "Truth and Probability." Ramsey's comments on this subject were very unsystematic, but I think that they advance considerations of fundamental importance which appear to have been overlooked by his followers. I may add that in Chapter III of Adams, 1975, I attempted to extend these ideas in order to provide a rational foundation for my theory of conditionals, and that argument has also been overlooked by commentators on this theory.[1] Thus, this paper bears both on the logic of decision and the logic of conditionals.

I will attempt to explain and outline the consequences of seven "theses", six of which are attributable to Ramsey, and the seventh of which involves a modification which is essential to my theory of conditionals. These are as follows.

THESIS 1. For systems of decisions or degrees of belief to be consistent is fundamentally a matter, not of their conforming to consistency defining laws, but rather of the possibility of all decisions or degrees of belief in these systems being *right* in a sense of rightness that is not logical but factual. In this, decision-theoretic consistency is analogous to the logical consistency of sets of propositions, which consists in the possibility of all propositions of these sets being true.

The problem of analysis is to make clear the sort of factual rightness that is appropriate to decisions and degrees of belief.

THESIS 2. Superficially, the rightness of degrees of belief consists in their corresponding to facts concerning class-ratios or relative frequencies.

This explains why consistent degrees of belief must conform to the laws of relative frequency, but it leaves us with the problem of explain-

William L. Harper and Brian Skyrms (eds.), Causation in Decision, Belief Change, and Statistics, II, 49—69.

ing which relative frequencies these are, and how it can be that degrees of belief purtaining to particular propositions are held to be right if they correspond to something general; namely statistics concerning class ratios. This also doesn't explain how the notions of rightness and consistency apply to decisions.

THESIS 3. More fundamentally, what makes either degrees of belief or decisions right is a pragmatic matter. Roughly, they are right if they are good or useful. A decision is good if the consequences of implementing it are what the decision maker aims at in arriving at it, and degrees of belief are good if the decisions that are based on them are good.

This conception, which is related to a pragmatic conception of truth, is able to explain why it is that we might *want* our degrees of belief and decisions to be right, and it also gives us something of a handle (but only a small one) on the problem of explaining why we should want to be consistent. What is still unexplained is the connection between this "pragmatic rightness" and relative frequencies, and in fact there appears to be a kind of conflict in connection with this. This has to do with the following:

THESIS 4. Degrees of belief that are ideally for the best are not relative frequencies. Ideal degrees of belief – those that lead most certainly to decisions with desired consequences – are always certainties, and in particular they must equal 1 for all true propositions and 0 for all false ones.

This conflict is resolved by:

THESIS 5. THE CENTRAL THESIS. It is a fact of human nature that persons are not always able to arrive at ideal degrees of belief, at least in circumstances in which they have to act. In many cases the best humanly attainable or *optimal* degrees of belief, judged in terms of the mean payoffs that result from acting on them, are proportions of truths in classes of "similar" propositions.

THESIS 6. The advantage that accrues from having and acting on degrees of belief that are optimal but not ideal has to do with the long run. Moreover, the only practical maxim that applies in the one-shot case is *be successful*.

THESIS 7 (Adams). In decision situations in which states and acts are not independent there is no applicable maxim that applies to the one-shot case, and mean long-run payoffs are optimized by acting on "conditional class ratios". These are the optimal degrees of belief applying to conditional propositions, and they conform to Jeffrey's "classical" and not "ratificationized" theory, though with an objectivist and long-run interpretation.

I will suggest at the end of this paper that these ideas have important bearings on currently controversial issues in decision theory including Newcomb's Paradox (and probability kinematics too, though that cannot be entered into here), though they don't resolve the controversies. But my main aim will be to explain the ideas as matters of interest in their own right.

2. THE *ΑΝΑΓΚΗ ΛΕΓΕΙΝ* AND *ΑΝΑΓΚΗ ΕΙΝΙΕ* OF LAWS OF BELIEF AND DECISION

Clearly decision theorists describe systems of decisions and degrees of belief as consistent or inconsistent, and they attempt to formulate laws to which consistent systems of these sorts must conform. They seem to advocate these laws as ones to which human beings *should* conform, presumably because some advantage is to be gained thereby. Thus, Ramsey says of the laws of degrees of belief that:

> ... a precise account of the nature of partial belief reveals that the laws of probability are laws of consistency, an extension to partial beliefs of formal logic, the logic of consistency. (Ramsey, 1978, "Truth and Probability", p. 84).

And Savage says in *The Foundations of Statistics*:

> ... the major use I would make of P1 and its successors is normative, to police my own decisions for consistency and, where possible, to make complicated decisions depend on simpler ones. (*Foundations of Statistics*, p. 20; P1 is the postulate that preferences should be simply ordered).

The question is: what does it mean to say that a system of decisions or of degrees of belief is consistent? Ramsey approaches this by considering the connection between this concept of consistency and the concept of consistency in ordinary truth-conditional logic, but this leads immediately to a problem. It seems that there is a fundamental dissimilarity between the two kinds of consistency, for while we can say that to believe both p and $-p$ would be inconsistent because they cannot both to be true, it seems that we cannot say that to attach degrees of belief of 0.9 to both of these propositions would be inconsistent because they cannot both be nine tenths true. Ramsey makes a similar point:

> This enables us to regard (not altogether correctly) formal logic as an objective science consisting of objectively necessary propositions. It thus gives us not merely the ἀνάγκη λεγειν that if we assert p we are bound in consistency to assert q also, but also the ἀνάγκη ειν that if p is true, so must q be. But when we extend formal logic to include partial beliefs this objective interpretation is lost; if we believe $p\&q$ to the extent of 1/3, and $p-q$ to the extent of 1/3, we are bound in consistency to believe $-p$ to the extent 1/3. This is the ἀνάγκη λεγειν but we cannot say that if $p\&q$ is 1/3 true and $p-q$ is 1/3 true, $-p$ must also be 1/3 true, for such a statement would be sheer nonsense. There is no corresponding ἀνάγκη ειν ("Truth and Probability", p. 89, my sentential notation).

Though he does not say so in so many words, I think that Ramsey regards the lack of an ἀνάγκη ειν or objective necessity corresponding to the ἀνάγκη λεγειν or necessity of thought ("epistemic necessity") in the laws of degrees of belief as a deficiency.[2] In any event he seeks to remedy the deficiency by describing an ἀνάγκη ειν that applies to the laws of partial belief.

> This is, however, possible in a roundabout way; we saw at the beginning of this essay that the calculus of probabilities can be interpreted in terms of class-ratios; . . . ("Truth and Probability", p. 89).

Before looking more closely at this idea, we should observe that Ramsey seeks something that corresponds to partial beliefs in the same way that truth does to ordinary "simple" beliefs. If this parallel is valid it ought to follow that what makes partial beliefs and decisions right, which is analogous to what makes simple beliefs true, cannot be established *a priori*. Rightness, whether of ordinary beliefs or of partial beliefs or decisions, is a matter of fact, not logic. Logic is concerned with the general laws to which right systems of beliefs or decisions conform. That is what consistency has to do with.

Now I want to consider Ramsey's ideas about the kind of rightness that attaches to partial beliefs. The superficial view, which has already been noted, is that what makes them right is their correspondence with class-ratios or relative frequencies.

3. THE CORRESPONDENCE BETWEEN PARTIAL BELIEF AND FREQUENCIES

That Ramsey holds that degrees of belief correspond in some sense to relative frequencies is evident:

> ... the very idea of partial belief involves reference to a hypothetical or ideal frequency. ("Truth and Probability", p. 90).

Assuming that what makes partial beliefs right is their correspondence with relative frequencies, we can ground the epistemic necessity of laws of partial belief in the objective necessity that attaches to the laws of these relative frequencies. To have degrees of belief equal to 0.9 in both p and $-p$ would be inconsistent because both of them cannot have relative frequencies equal to 0.9. Thus, shortly after the passage just cited Ramsey says:

> It is this connection between partial belief and frequency which enables us to use the calculus of frequencies as a calculus of *consistent* partial belief. ("Truth and Probability", p. 90, my italics).

But this is superficial, because it is not clear how to attach relative frequencies to particular propositions. More important, it is not clear what advantage there might be in having degrees of belief that are "right" in the sense that they correspond to these frequencies, which might in turn explain what the advantage is of having consistent degrees of belief. As I already said, I think that Ramsey's more fundamental idea is that what stands to degrees of belief and to decisions as truth does to ordinary propositions is a pragmatic value, and more specifically it is that of being the best degrees of belief and decisions that human beings are capable of attaining. But clarifying this claim requires that we first make a detour and consider "ideal" degrees of belief, which would be even better. I will now try to explain this, though here I will be going out on a limb – perhaps more than one limb – in interpreting Ramsey.

4. IDEAL DEGREES OF BELIEF

The following quotation describes the degrees of belief that would be "ideally for the best":

> ... ideally the best thing is that we should have beliefs of degree 1 in all true propositions and beliefs of degree 0 in all false propositions. ("Truth and Probability", p. 86).

To see the grounds for this claim, consider the decision matrix in Figure 1 with two states (corresponding to the propositions p and $-p$) and two acts, and with degrees of belief x and $1-x$ and utilities a, b, c, and d:

	p Deg = x	−p Deg = 1−x	exp. utility
Act 1	U = a	U = b	xa + (1−x) b
Act 2	U = c	U = d	xc + (1−x) d

Fig. 1.

Assuming that this is not a dominance situation the utility maximizing act (the Bayes act) depends on x and $1-x$, and whether the act chosen is the "right" or "best" one depends on the actual state of affairs. For instance, if $a=d=1$ and $b=c=0$ Act 1 will be the Bayes act if $x > 0.5$, and that will be the right choice if p is actually the case. Under the circumstances we can say that the decision maker's degrees of belief are right if $x > 0.5$. But this isn't necessarily ideal, since if a, b, and c remain the same but d is increased to 99 Act 1 will still be the right choice if the actual state is p, but it won't be chosen unless $x >$ 0.99. Clearly ideally the best degrees of belief, which are those that lead infallibly to the right choices, will attach 1 to actual states of affairs and 0 to all others, though it is also clear that less-than-ideal degrees can be "good enough" under certain circumstances. Also, reiterating a point made earlier, it is clear that whatever these less-than-ideal but good enough degrees of belief are, they must depend on facts that are determined *a posteriori*, and they will not be matters of pure reason. The problem is to describe less-than-ideal but nevertheless humanly attainable degrees of belief and to relate them to relative frequencies in the way that Ramsey hints is possible. This will be discussed in the following section, but a couple of parenthetical remarks are in order before coming to that.

The argument that ideal degrees of belief are equal to 1 for true propositions and 0 for false ones presupposed the principles on which the above matrix representation was based, which are essentially those of L. J. Savage's decision theory. But Savage's theory, and especially its Sure Thing Principle, presupposes that states are independent of acts. When states are influenced by acts we have to modify the decision-theoretic picture, and consider generalizations such as those of Jeffrey, 1965 and 1984, or of Harper and Gibbard, 1976. The latter theory involves the degrees of belief that should be associated with "act-conditionals" like "If the decision maker performs Act 1 then consequence *C* will ensue" or perhaps "If the decision maker were to perform Act 1 consequence *C* would ensue." These are at the heart of current controversies relating to Newcomb's Paradox, but they also involve my theory of conditionals because I hold that my theory applies to act-conditionals. This will be returned to Section 8.

The second point is that the pragmatic notion of rightness of degrees of belief goes hand in hand with the idea of decisions or choices being right. Act 1 is the right choice in the decision situation represented in the matrix above, given that $a=1$ and $b=0$ and that p is true, because that is the act that leads to the most desired consequence in this situation. Thus we have a notion of rightness that applies to decisions, and therefore an objective characterization of consistency that applies to a system of decisons: namely that it should be possible for all decisions of the system to be right. I think that it is fundamentally in terms of this notion of consistency that claims, e.g. that consistent decisions can violate the Sure Thing Principle, must be justified. As it happens these violations arise in the act-dependent case, the discussion of which will be postponed to Section 8.

We will now turn to the problem of explaining in what sense degrees of belief other than 0 or 1 can be pragmatically right, even though ideally right degrees of belief are always certainties. This has to do with the fact that human beings are not always able to be ideally right, or as Ramsey says immediately following the passage just quoted:

> ... this is too high a standard to expect of mortal man, and we must expect that some degree of doubt or even of error may be humanly speaking justified. ("Truth and Probability", p. 86).

5. THE BEST HUMANLY ATTAINABLE DEGREES OF BELIEF

We should be clear that the ideal of attaching degree of belief 1 to truths can be attained in any particular case, and even in any particular situation. For instance, it is obviously possible for a person guessing the fall of a coin that is about to be tossed to be irrationally certain that it will fall heads, and for the person then to be "vindicated" by the coin's actually falling heads. But while persons may be able to do this on particular occasions we know that they cannot do it all of the time. To quote Ramsey again,

> . . . the highest ideal would be *always* to have a true opinion and be certain of it; but this ideal is better suited to God than to man. ("Truth and Probability", pp. 95–6, my italics).

That men cannot always hold true opinions and be certain of them is something we know not *a priori* but from ages-long experience of human fallibility. This experience gives us statistical knowledge relating to human powers of prediction, and I am now going to suggest that *optimal* degrees of belief, which are the best that human beings are capable of attaining, correspond to statistics of this sort. The correspondence is hinted at in the following quotation:

> . . . belief of degree m/n is the sort of belief which leads to action which would be best if repeated n times in m of which the proposition is true; . . . ("Truth and Probability", p. 90).

This is inexact as it stands since a proposition cannot be true at one time and false at another,[3] but it can be interpreted as meaning that among constant or *uniform* degrees of belief that could be attached to propositions of a class, the best is that which equals the proportion of true propositions in the class.[4] This degree would be best because it would lead to actions whose *mean payoffs* are best among all sequences of actions to which uniform belief-degrees would lead. Of course, in the case of classes containing just one proposition the rule that best uniform degrees of belief equal proportions of propositions that are true reduces to the rule that ideal degrees of belief equal 1 for true propositions and 0 for false ones. But, whereas it may not always be possible to attach ideal degrees of belief to individual propositions, in certain cases it may be possible to attach best uniform degrees of belief to certain classes to which they belong.

A particular case is that of classes of *all propositions of a given kind*, and the belief-degrees that might be attached to them in *recognizable situations*. For example the propositions might be ones expressible as "the coin will fall heads" in reference to a coin that is being tossed, and the degrees of belief in question might be those that could be attached to these propositions prior to the fall of the coin.[5] While persons may be unable to recognize that subclass of situations in which the coin ultimately falls heads, they can at least recognize that the coin is being tossed, and given this they can be assumed capable of following a policy of attaching a uniform belief-degree to the propositions in these situations. If this degree equals the proportion of times the coin actually falls heads it will be the best uniform belief-degree that could attach to propositions of this class. That is not to say that these are the best belief-degrees that human beings may be capable of assigning, since there might be even better non-uniform policies that they could follow in these situations.

There is a special case in which no superior non-uniform policy could be followed. That is that in which the degree of belief are attached to propositions that are *psychologically random* in the situations in which the beliefs attach (cf. Reichenbach, 1949, p. 150), in the sense that however human beings select among these situations ones in which to predict that the propositions are true, they will be right in these predictions only a given proportion of the time in the long run. For instance, to a first approximation it is a fact of experience that no matter how people in situations in which coins are being tossed choose whether or not to predict that the coin will fall heads, they will be right only 50% of the time in the long run. Propositions expressed by "the coin will fall heads" are therefore psychologically random in these situations, and this implies that no humanly feasible non-uniform policy of assigning degrees of belief to these propositions in these situations can lead to actions with better mean payoffs in the long run than the uniform policy of assigning them all a degree of belief of 0.5.

Roughly generalizing, it can be argued that where actions are taken in situations in which the degrees of belief acted on apply to propositions that are psychologically random, persons will gain the maximum humanly attainable mean utility payoffs in the long run if their degrees of belief equal the proportions of propositions which are true in the situations. These are the *optimal* degrees of belief in the given situations. In its essentials this is the argument that was given in Section III.4 of Adams, 1975, though it cannot be gone into in more detail here.[6] Implications will be considered below, but a few incidental remarks will be made first.

First: while correct prediction with certainty may be something only a diety is capable of, it is possible for human beings to assign optimal degrees of belief to propositions of classes that are psychologically random relative to recognizable situations. But that is not to say that all degrees of belief are optimal, or that it is possible for persons to know that their degrees of belief are optimal. Approximating the optimal may be regarded as a goal of rational inquiry, in a sense of rationality that goes beyond mere consistency.

Second: my interpretation of "right" or "optimal" degrees of belief makes these degrees depend not only on the propositions but on the situation in which they are considered. Describing the way in which optimal degrees depend on situations is one interpretation of the problem of probability kinematics, though details cannot be entered into here.

Third: while it is in particular situations and with respect to particular propositions that certain degrees of belief are optimal, nonetheless what makes them optimal is their agreement or disagreement with proportions of "similar" propositions which are true. This explains why optimal degrees of belief satisfy the laws of the calculus of frequencies, and therefore these are the laws to which consistent degrees of belief conform.

Fourth: while I have argued that the optimal degree of belief to attach to a proposition in a situation equals the proportion of truths in a class to which it belongs, the problem of ambiguity of reference class does not arise. That is because if a proposition p is considered in a situation s, where p belongs to two proposition classes P and P' and s belongs to two infinite classes of recognizable situations, S and S', where members of P are psychologically random in situations of class S and members of P' are psychologically random in situations of class S', then P and P' must contian the same proportions of true propositions.

The final observation is that it is not the case that optimal degrees of belief of the sort I have been describing exist for all propositions, p, in all situations, s, because there may be *no* classes S and P to which s and p belong such the propositions in P are psychologically random in situations in S.[7] In particular I would argue these don't exist in situations in which one is in a position to influence what is being predicted, which is typically the case with the antecedents and consequents of the act-conditionals. However in that case there may be still be a kind of *conditional* psychological randomness, and the associated conditional truth proportions still furnish the basis for optimal decision-making in such situations. Section III.7 of Adams, 1975, argues that these frequencies are the optimal degrees of belief to attach to act-conditionals, and it will be suggested in Section 8 below that these are the most advantageous probabilities on which to base decisions in the Newcomb situation.

Now we will turn to implications of the views just sketched.

6. THE ADVANTAGES OF CONSISTENCY

I think that Ramsey took seriously the too often ignored question of why persons should attempt to be consistent, but what he said about it was equivocal:

> We put before ourselves the standard of consistency and construct these elaborate rules to insure its observance. But this is obviously not enough; we want our beliefs to be consistent not merely with one another but with the facts; nor is it even clear that consistency is always advantageous; it may well be better to be sometimes right than never right. ("Truth and Probability", p. 93).

On the one hand it seems that we *aim* to be consistent, presumably because some advantage attaches to that, but on the other it seems that consistency may not be advantageous after all. It isn't easy to sort this out, and here I must digress from Ramsey's views and make some suggestions of my own.

I am inclined to doubt that we do "put before ourselves" standards of consistency, at least with the aim of securing direct practical advantage. As Ramsey says, rightness is what is advantageous practically, and consistency does not guarantee that. I would go even farther. Though I can't prove this, I suspect that the expected utility of actions based on randomly chosen consistent systems of degrees of belief is not significantly higher than that of actions based on *any* random degrees of belief, consistent or not. If this is so then the "practical marginal utility" of being consistent is negligible.[8] What then is the status of consistency in practical reason and decision? I will hazard a couple of speculations.

One is that the techniques of inquiry we use to arrive at right degrees of belief are likely to lead to convergence to consistency, simply because they converge to rightness and right beliefs are consistent. Certain of these techniques may involve noting inconsistencies as a sign that more inquiry is required. This would give a probabilistic interpretation to a claim of Quine and Ullian's about the "probing" that persons undertake as a result of discovering that their beliefs are in conflict:

> In practice, the probing stops when we are best satisfied how to restore consistency. (*The Web of Belief*, p. 10).

Another speculation is that the most immediate practical utility of consistency may relate to communication. It is possible that a person could have random success basing his betting behavior on degrees of belief equal to 0.9, both that a coin will fall heads and that it will fall tails, but he would only create puzzlement if he *said* that these are what the odds were.

The third point is that consistency is not all there is to rationality. As already noted there are techniques of inquiry for arriving at degrees of belief that are right or advantageous, and these techniques are rational, if at all, in a sense that transcends mere consistency. Whenever there is dispute over the rationality of particular choices or degrees of belief, as against systems of them, it is plausible that this concerns not consistency but some sort of rightness. Since much of the current controversy relating to Newcomb's Paradox concerns a particular choice, the problem there is to make clear what sort of rightness is in dispute, and what its advantages may be. This leads to a very general observation about both consistency and rightness.

Assuming that a system of beliefs or decisions is consistent if it is possible for all decisions or beliefs in the system to be right, and rightness consists in being good or useful, consistency is relative to what is held to be good. Furthermore what is held to be good must depend on the decision maker, and it is possible that the same is true of the general laws these that these "goods" conform to. Comments in the following section on the long and short run suggest that this is a real possibility, and it could be that some of the disagreements about the Newcomb Problem arise from different conceptions as to what should motivate decision makers in this situation.

Whether or not these speculations are right, they seem to me to point to the conclusion that consistency is not an end-in-itself, and I would say the same of rationality in general. In consequence if certain laws are held up as normative standards of consistency, and if they are justified at all, this must be because conformity to them serves purposes beyond securing "consistency for its own sake". Hence, if there is dispute concerning what is or is not consistent or rational, it must remain sterile so long as the advantages of consistency or rationality are not made clear.

7. THE SHORT-RUN AND THE LONG-RUN

I said that, given that long run gains are sought, pragmatically best degrees of belief are proportions of truths in sets of propositions that are psychologically random in the situations in which they are considered. But, as Keynes said, "in the long run we'll all be dead," and it is doubtful that persons really are motivated by the desire for gains that are measured in terms of an infinitely long run. The question is: are we justified in recommending conformity to principles of rationality that are designed to secure long run benefits to persons who do not desire them? The following highly speculative remarks will have important consequences so far as the Newcomb Problem is concerned.

First, it is obvious that there is no guarantee that taking actions or following policies that are best calculated to yield long run advantage will yield short-run advantages. The gambler who stakes his all that a coin will not fall heads on 10 successive tosses may end by losing all. Nor can we argue that we should act in the short run as though we sought long run gain, because doing so will yield short run gains most of the time. To insure short-run gain most of the time is not to insure it at any particular time, which is what is sought in the one-shot case. Given this, it seems to me to be an unjustified "transcendental" assumption to suppose that persons who aim solely for success in the single case should act as though they sought long term average gains.

But in real life actions that are properly described as aiming just for success in the single case are probably very rare, and often ones that seem to aim for this actually have longer range goals. These may include acts in accord with long run policies which direct that action be taken in any particular case as though that case were an end in itself. For instance, to a first approximation it is a reasonable policy for a baseball team aiming to win the pennant to play each game as though winning that game were all that mattered.[9] But to play in this way is not for it to be the case that winning the individual game *is* all that matters. And, it would be wrong to regard playing as though the single game were all that mattered as acting in accord with the transcendental principle that when the single game is all that matters the rational thing is to act as though success in the long run were being sought.

So far as I can see the only practical maxim for the one-shot case is just *be right* (or at least *be successful*), and all other practical maxims purtain to the long run. Ramsey suggests something like this when he says:

> Thus given a single opinion, we can only praise or blame it on the ground of truth or falsity; ... ("Truth and Probability", p. 98).

Coupling this with the dictum that "... the very idea of partial belief involves reference to a hypothetical or ideal frequency" suggests the following. That is that apparent disagreements about what is rational to do or think in the one-shot case, which involve considerations of degrees of belief, must have other than immediate practical values in mind, such as that of intelligable communication, or they mistake cases of acting as though the single case is all that matters for ones in which it *is* all that matters.

The above is an admittedly debatable interpretation of Ramsey, which must have radical consequences for decision theory. Among these is its generalization to the act-dependent case and the Newcomb Paradox, which will be commented on very briefly in the concluding section.

8. THE ACT-DEPENDENT CASE

It is not possible to enter into the same degree of detail in discussing this case as it was in the act-independent one, but a few comments will serve to suggest the modifications of the latter that are required. The first thing to note is that the thesis that pragmatically ideal degrees of belief equal 1 for true propositions and 0 for false ones cannot be maintained in the act-dependent case. To see this reconsider the decision matrix in Figure 1, but this time applied to the case in which the possible states are that the decision-maker gets cancer or that he doesn't, the acts are to smoke or not to smoke, and suppose that the utilities of the possible consequences (in some unspecified units) are as shown:

	get cancer Deg = x	don't get cancer Deg = 1−x	exp. utility
smoke	U = −1000	U = 500	500 − 1500x
don't smoke	U = −1500	U = 0	−1500x

Fig. 2.

This is a dominence situation, hence whatever *x* is the utility maximizing act will be to smoke. Furthermore, according to the reasoning of Section 4 this is the right choice whichever state is the actual one.

But we can no longer claim that attaching degree of belief 1 to whichever state proves to be the actual one and acting accordingly is for the best. If the decision maker is fatalistically certain that he will get cancer and smokes and does get cancer he will have "held a true opinion and been certain of it", but that would not have been for the best if the smoking was what actually caused the cancer.

Given that it is not ideally for the best to hold true opinions and be certain of them about matters that these opinions influence, or at least that the argument given for this in Section 4 no longer applies, it is implausible that optimal degrees of belief concerning these matters can be equated with proportions of propositions that are true. But it is significant that the propositions in question are not psychologically random in the act-dependent situation. This is most clearly seen in the case of propositions which assert that a decision maker will make a particular choice, considered by the decision maker himself in the situation in which he is still deciding what choice to make. Thus, suppose that a person deciding whether to toss a coin considers the proposition "I will toss the coin." This is the sort of recognizable situation in which the person can influence the chances of being right simply by choosing to confirm or infirm his own predictions, and evidently judgements about this type of proposition are not psychologically random in these situations.

The foregoing suggests that defining pragmatically best decisions by generalizing from ideal one-shot beliefs may be the wrong approach to take in the act-dependent case. My own approach is to avoid entirely attempting to formulate principles that apply to this case, and instead consider directly what is best in the long run. Here we may be helped by a key observation. That is that there are many situations which are not unconditionally psychologically random but which are *conditionally* so, in which optimal conditional frequencies may be defined. For instance, in the coin-tossing case, while the tossing of the coin may not be psychologically random in the situation in which the decision maker is making up his mind whether to toss it, the coin's falling heads will be psychologically random conditional on the coin's being tossed. More exactly, no matter how the decision maker chooses to make predictions of the form "the coin will fall heads" in these situations he will be right 50% of the time in the long run in those cases in which he actually tosses the coin. This applies to decision making in the following way.

Return once more to the decision matrix of Figure 1, but now suppose that the decision maker is choosing between Acts 1 and 2 in a situation in which *p* and −*p* are psychologically random conditional on these acts being performed. Here, instead of writing the unconditional frequencies along with the propositions, the conditional frequencies (abbr. "cf") are written into the cells of the matrix along with the utilities:

	p	-p	exp. utility
Act 1	cf = x U = a	cf = 1-x U = b	xa + (1-x)b
Act 2	cf = y U = c	cf = 1-y U = d	yc + (1-y)d

Fig. 3.

It is argued in Adams, 1975, Section III.7 that the decision maker will have an optimum mean payoff in the long run if he always chooses the act with the maximum expected utility as given by the matrix above. This means that he has the exactly the same reason to make this choice as he had for choosing the utility maximizing option given by the matrix in Figure 1 in the act-independent case.

There are several remarks to make in closing. First, the rule to choose the act with the maximum expected utility as given by the above matrix is formally the same as the decision rule of the "classical" and not "ratificationized" Jeffrey theory (i.e., the theory of Jeffrey, 1965, and not of Jeffrey, 1983). There are important differences of interpretation, however. One is that this rule attempts to describe decisions which are optimal in the sense that they yield mean payoffs in the long run that are the highest that human beings can attain. It is secondary that decisions made in accord with this rule must be consistent, and it is not argued that it is advantageous for decision makers to choose on the basis of "mistaken" conditional frequencies which are nevertheless consistent with the formalism.

Another important difference is that in bypassing the single case I appear to have bypassed degrees of belief as well. The conditional frequencies entering into the matrix of Figure 3 are not optimal degrees of belief attaching to the unconditional propositions involved, which I have argued do not exist in the act-dependent situation. *Ipso facto* they are not ratios of such degrees of belief. It is a further step, which is crucial to my theory of conditionals, to argue that these frequencies are optimal degrees of belief to attach to associated *conditional* propositions.[10]

A third difference is that the present theory is neither epistemic nor explicitly causal. Whether or not certain decisions are optimal has nothing to do with the "welcomeness of news", whatever that is, and, while optimality depends on the decision-making situation it has nothing to do with what the decision maker knows about it. It is true that we may ultimately attempt to explain act-dependence in causal terms, but my description of conditional psychological randomness is a quasi-modal one having to do with limitations on human powers of decision-making. Furthermore, given this sort of randomness the conditional frequencies entering into the optimal decision are matters of empirical fact and it is a purely mathematical fact that acting on these frequencies will yield maximum long run gains.

For present purposes perhaps the most important difference in interpretation is that this analysis concerns the long-run. Nothing is said about what it is rational to do in situations in which the single case is all that matters, whereas the Jeffrey theories (as well as those of Eells, 1982; Harper and Gibbard, 1978; Lewis, 1981; Skyrms, 1984, Chapter 4; and Sobel, 1983, to mention just a few) purport to apply to that case. That is not to say that the latter theories are wrong. However, I feel that at present they are fundamentally deficient in failing to provide an account of the advantages that can be expected to follow from adhering to their tenets.[11]

It is the last observation that bears most directly on the Newcomb Paradox. In this case we can suppose that the payoff matrix is that given in Figure 4:

	\$1000 in opaque box	nothing in opaque box	exp. utility
take both boxes	cf = x U = 1010	cf = 1−x U = 10	1000x + 10
take opaque box	cf = y U = 1000	cf = 1−y U = o	1000y

Fig. 4.

Calculating here on the basis of classical Jeffrey utilities should lead the decision maker to take the second alternative (i.e., he should only take the money in the opaque box) if $y > x+0.01$, but the question we have to ask is whether that is the right decision. Recall that we are not concerned with the consistency of a set of decisions, and moreover something like degrees of belief are in the picture (that would be the case if the conditional frequencies in the cells of the matrix were ratios of unconditional belief degrees, as Jeffrey holds) and so it is long run advantage that it is in question.

The usual formulations of the Newcomb Paradox do not make it clear whether the "consequence propositions" in the decision matrix (that there is or there is not $1000 in the opaque box) are conditionally psychologically random in the decision situation, but let us suppose so: i.e., suppose that no matter how the decision maker varies his choices the consequences of choosing Act 1 will occur in fixed proportions x and $1-x$, and the consequences of choosing Act 2 will occur in proportions y and $1-y$. Given this, the choice which yields the maximum mean payoff in the long run will be Act 2 if $y > x + 0.01$. That is simply an arithmetical truth, and the only question that arises has to do with its significance.

If maximum long run payoff is what is sought and $y > x+0.01$ then Act 2 will be the right choice, whether the decision maker knows it or not. What if the decision maker does not seek maximum long run payoff?. As already stated, the present theory offers no maxims that apply to the one-shot case, and if Ramsey is right there are no useful ones to apply beyond "be right" – and in the act-dependent case it is not even a clear what that is. What is it rational for the decision maker to do if he does seek maximum long run payoff but he doesn't know what the right choice is because he doesn't know the relevant conditional frequencies? Then the problem of rationally estimating them arises. There again the present theory offers no precepts, though at least it is clear what has to be estimated. And, that is as far as I have been able to go.

University of California, Berkeley

NOTES

[1] Actually, at the time of writing my book I was not aware of the extent to which Ramsey and anticipated me, and consequently I failed to acknowledge his precedence.

[2] I am indebted to Alan Code for enlightenment concerning the Greek terms, but I have chosen to render ἀνάγκη λεγειν as *necessity of thought* following the parallel with Ramsey's contrast between *ratio cognoscendi* and *ratio essendi* in "General Propositions and Causality", p. 145 (my attention was drawn to this by John Vickers).

Though it is off the present topic, I cannot refrain from commenting on another matter related to "General Propositions and Causality". This is the *locus classicus* of the widely commented on *Ramsey Test* analysis of conditionals ("General Propositions and Causality", footnote, p. 143), which some writers have claimed is my own view (cf., Harper, 1981, p. 14). This is erronious. The reader will discover on reading further in this paper that my theory concerns the *optimal* degrees of belief that should attach to conditionals under various circumstances, and that has nothing at all to do with the mental operations which the thinker may perform in arriving at these beliefs (e.g., imagining adding propositions to his stock of knowledge, as in the Ramsey Test).

[3] Wesley Salmon suggested in oral comments on this paper that Ramsey's claim might be explicated in terms of *propositional functions* rather than propositions. In fact this would fit in with an immediately prior claim of Ramsey's:

> . . . a partial belief cannot in general be connected uniquely with any actual frequency, for the connection is always made by taking the proposition in question as an instance of a propositional function. ("Truth and Probability", p. 90).

It would be interesting to develop this idea systematically.

[4] Note also in this connection Ramsey's remark about the best degree of belief for a person to have concerning the unwholesomeness of yellow toadstools:

> . . . given that he is going to think always in the same way about all yellow toadstools, we can ask what degree of confidence it would be best for him to have that they are unwholesome. And the answer will be that it would be best for his degree of belief that a yellow toadstool is unwholesome to be equal to the proportion of yellow toadstools that are in fact unwholesome. (This follows from the meaning of degree of belief). ("Truth and Probability", p. 97).

[5] The present argument is informal, and making it more precise would involve introducing of the idea of classes of situation-proposition *pairs*, $\langle s,p \rangle$, in which s is a situation and p is a proposition standing in a specified relation to s. An example would be a class of pairs in which s is a "coin-tossing situation" and p is a proposition expressible in that situation by saying "the coin will fall heads". There are further inaccuracies in the present argument which cannot be discussed in this paper.

[6] Actually, I argue for something slightly stronger in Chapter III of my book. That is that persons seeking maximum mean payoffs in the long run are best off if they *act as though* they choose actions with maximum expected utility, and in so doing they act as though they base these choices on degrees of belief that are right in the sense that they correspond to class ratios. This argument cannot be given here, but I hold that something similar applies in the act-dependent case of which Newcomb's Paradox is an example.

[7] It must be admitted that in allowing that there may not be "correct" degrees of belief corresponding to all propositions I may be going against Ramsey's own views. As Isaac Levi pointed out in oral comments, Ramsey's springboard in "Truth and Probability" for the development of his constructive views was a critical commentary on Keynes's theory of probability, including the tenet that numerical probabilities do not attach to all

propositions. For the present I do not see how to reconcile the thesis that optimal degrees of belief attach to all propositions with the thesis that these degrees are determined by statistics concerning limitations on human judgemental capabilities.

[8] Obviously this runs counter to the arguments of the advocates of Dutch Book justifications for the laws of degrees of belief. My objections to this approach are set forth in Section 2 of Adams and Rosenkrantz, 1980.

[9] The pennant race example suggests that policy will direct that the single case should be treated as though it were all that mattered when long-run success (winning the pennant) derives from short-run "incremental gains" (winning individual games). To what extent which this is true of long run policies in general is something that should be examined.

[10] Actually, no *argument* is given for this in Adams, 1975. More recent work, as yet unpublished, attempts to approach this issue from the analysis of the "information content" of speech acts in which speakers make assertions of conditional form such as "If you smoke you will get lung cancer." Adams, 1986, discusses this matter informally.

[11] Examples of unsubstantiated claims that adherence to certain principles is "bad" and adherence to others is better abound in the literature. Thus Jeffrey says, commenting on the maxim of expected utility maximization. "But in the prisoner's dilemma, and in other, more far-fetched cases such as Newcomb's problem, that maxim produces bad advice" (Jeffrey, 1984, p. 10). We are given no clear indication as to why following this advice would be bad, or why a decision maker would be better off if he acted on the basis of *ratifiable* degrees of belief.

REFERENCES

Adams, E. W. (1975) *The Logic of Conditionals, an Application of Probability to Deductive Logic*, D. Reidel, Dordrecht.

Adams, E. W. 'The semantics and pragmatics of conditionals', to appear in Traugott, *et al.* (eds.), *On Conditionals*, Cambridge University Press.

Adams, E. W. and Rosenkrantz, R. (1980) 'Applying the Jeffrey decision model to rational betting and information acquisition', *Theory and Decision* **12**, 1–20.

Eells, E. (1982) *Rational Decision and Causality*, Cambridge University Press.

Gibbard, A. and Harper, W. L. (1978) 'Counterfactuals and two kinds of expected utility', in C. A. Hooker, J. J. Leach, and E. F. McClennen (eds.), *Foundations and Applications of Decision Theory*, Vol. I., D. Reidel, Dordrecht, 153–190.

Harper, W. L. (1981) 'A sketch of some recent developments in the theory of conditionals', in W. L. Harper, R. Stalnaker, and G. Pearce (eds.), *Ifs*, D. Reidel, Dordrecht, 3–38.

Jeffrey, R. C. (1965) *The Logic of Decision*, New York, McGraw-Hill.

Jeffrey, R. C. (1983) *The Logic of Decision, 2nd edition*, Chicago, The University of Chicago Press.

Lewis, D. (1981) 'Causal decision theory.' *Australasian Journal of Philosophy* **59**, 5–30.

Quine, W. V. and Ullian, J. (1970) *The Web of Belief*, New York: Random House.

Ramsey, F. P. (1978) *FOUNDATIONS, Essays in Philosophy, Logic, Mathematics and Economics*, edited by D. H. Mellor. Atlantic Highlands, Humanities Press.

Reichenbach, H. (1949) *The Theory of Probability*, *2nd edition*, English translation by E. H. Hutten and M. Reichenbach. Berkeley: University of California Press.

Savage, L. J. (1954) *The Foundations of Statistics*, New York, John Wiley and Sons.

Skyrms, B. (1984) *Pragmatics and Empiricism*, New Haven: Yale University Press.

Sobel, H. (1983) 'Expected utilities and rational actions and choices', *Theoria*.

HENRY E. KYBURG, JR.

POWERS

1

I have two quite distinct themes I would like to bring together here. One concerns the metaphysical status of causality, and one way to get at that is to look closely at the role of causal principles in classical mechanics. The other concerns the role of causal relations in analysing choice under certainty, where the uncertainty may appear to be of two sorts – causal uncertainty, or epistemic uncertainty. The latter sort of problem has attracted a lot of attention and has motivated a variety of notions of utility, of probability, of choice principles, and so on.

The two themes are related. With regard to classical mechanics, I conclude that causality plays no role whatever, except possibly an obscurely psychological one in the motivations of physicists. Metaphysically speaking, causality is sheer superstition. Having purged ourselves of the superstition, we then look afresh at those tricky decision problems in which causality seems to play a role: the decision problems that seem to separate causal decision theorists from epistemic theorists. We find that there is indeed a sharp distinction to be drawn between two kinds of analysis, but that (1) the epistemic/causal distinction misses the point, (2) the distinction can be captured in almost classical terms, and (3) thus characterized, the distinction is a phenomenologically familiar one.

2

Classical physics is deterministic in the sense that, given the positions and velocities of an isolated set of particles, together with the forces acting on them, the state of the system at any future time is fixed. Or, for that matter, any past time. This very symmetry makes the association of causality with this determinism suspect. To capture some of the directional idea of causality, we need to think of the present state of the system as having the *power* to produce the future state, but not vice versa. Less anthropomorphically, I can explain the present state in

William L. Harper and Brian Skyrms (eds.), Causation in Decision, Belief Change, and Statistics, II, 71–82.

terms of the past state, and the laws of physics; I can predict the future state, given the laws of nature and the present state. Think of La Place's demon – he's just like us, but knows more.

It has been suggested that what distinguishes classical physics from quantum mechanics is that in classical physics you can measure the state of the system with arbitrary precision at any point on its trajectory through state space, and predict its state at any other point with arbitrary accuracy. But this is odd – classical physics doesn't talk about *measurement* at all, if we think of it as consisting of Newton's laws and some force functions. Classical physics is about how certain *quantities* are related.

If we are going to make any such claim about classical and modern physics, we had better think about measurement in classical physics, and therefore also about the theory of errors of measurement. I claim (and have tried to show elsewhere [1]) that we can and do get our theory of errors of measurement directly through statistical inference. The story isn't simple, because it is tied to the simultaneous development of theory, but to get the flavor, consider the measurement of the length of an object. Assume that you can identify the object, and that it doesn't change length, and you have a number of measurements of the same quantity. Take those measurements to be a sample from a hypothetical population of measurements of that quantity, and you can make a statistical inference that says that it is highly probable that the measurements have approximately such and such a distribution in the hypothetical population. Together with some pragmatically plausible principles, you can get at error distributions in this way.

But what is important to bear in mind is that – as in the case of statistical inference in general – the *deductive* output of these inferences is that the probability is high that the errors of measurement of a certain sort are distributed approximately according to a certain sort of distribution. Thus: the probability is 0.99 that errors of measurement of length with this procedure are approximately normal, with a mean close to 0 and a variance of about 0.01.

To use this sort of knowledge, we have to indulge in an inductive leap: we have to say that the distribution of errors *is* approximately normal with a mean close to 0 and a variance of about 0.01. And now we make a measurement: we read *m* on the scale, or see *m* displayed on the screen. What can we say about the value of the underlying quantity? We can say – taking quite a few things for granted, but this is the way responsible people seem to talk – that the probability is 0.99 that the underlying quantity lies in the interval $m \pm d$. But that does not yield any categorical predictions about the future values of the quantity. For that sort of thing, we need another inductive leap: we have to say that the quantity in question *does* lie in the interval $m \pm d$.

Where do we get this talk of "arbitrary precision", then? Generally speaking, we can make instruments more accurate than those we happen to have right now. The theory of error that we have derived statistically from the examination of measurements generally takes the errors of several measurements of the same quantity to be (approximately) independent, so that the variance of the error of the average of n of them will be reduced by a factor of $1/n$, for reasonable n. Furthermore, in the interesting cases, we are talking of the measurement of a number of different quantities, and we suppose (on good statistical grounds) that the errors of measurement of these different quantities are approximately independent.

La Place's image was inspired by astronomical successes. We can predict where Venus will be at a certain time in the future. Perhaps La Place thought of gravitational attraction as a power producing that result. On closer examination, we see that such predictions are shot through with inductive acceptance, statistical approximation, and small uncertainty, despite the fact that the classical view is undeniably deterministic.

What we had excellent grounds for believing was that the theory worked to a good approximation. We had no grounds whatever for thinking that in the arbitrary limit, with unbounded replications of measurements, it would work exactly. But this is what is necessary for the clockwork causal universe. (More accurately, what is necessary is that the deviations decrease stochastically over time; but that is not how the laws of our universe work. A small error in the mass of Venus leads to an *increasing*, not decreasing, error in the date of its successive apogees.)

Or consider a dead leaf falling from the top of an oak tree on a slightly breezy fall day. Its center of mass (presumably) follows a certain trajectory in falling to the ground. That trajectory is (on the classical view, and few would invoke quantum mechanical uncertainties in this context) determined by the mass distribution and shape of the leaf, the moments imparted to it at the moment of its detachment from the tree, the velocity and viscosity and distribution of the air stream moving past it, and so on. The most ardent believer in the sanctity of causality in classical physics would not maintain that we can predict this trajectory – no matter what measurements we made. True, given (hypothetically) initial conditions, there *is* a determined trajectory according to classical physics. But this is a consequence of classical physics that is not only untested and untestable, but uninteresting because so completely out of our control and empty of influence on us.

Why do we associate causality with classical physics, then? In part in may be that it is because we can do a good job of predicting certain sorts of things (the next high tide); but in greater part I think it is because classical physics gives us the power to manipulate events in the world. We can *make* things happen. What happens is the result of a decision, an action, on our part. I push the button, which closes the circuit, which activates the radio transmitter, which transmits electromagnetic emissions, which are received by a receiver, which sends an impulse down a wire, which activates a detonator, which explodes, setting off a charge of TNT, which shatters a rock ledge. I may have intended to help with the building of a road, or I may have actually intended to turn on the television set. No matter. The sequence of events begins with an *act*. Of course we can and do break up the sequence into *uniformities* – hence "sequence" – but the power that is transmitted along the sequence is the power of an agent.

Causality is in the first place a notion associated with action – it is *agents* who cause things – and only derivatively, I suggest, a notion associated with inanimate objects. One can see the transition. I can make a fire. My enemy can cause my house to fall down. But some things are beyond my control (I can't exercise power over them). I can't kill a bear single handedly; I need the help of my clan. Some things are beyond our control collectively – the weather, the course of a pestilence – so we attribute them to the *actions* of a deity.

At the same time, we try to bring these things (drought, pestilence) back under control, by getting control over the deity by prayers and supplications – that is, bringing these large events back under our *power*. This is superstition. There is also the fact that we can extend our power over things by taking advantage of certain sorts of uniformities, both statistical and universal. I can *make* my sword harder by tempering it (taking advantage of a uniform relation between heat treating and hardness); I can *make* my sweet peas more uniform by allowing only the red ones to pollinate. This is science.

Note that with science we may not have got rid of the demons that ruled the primitive world. We may just have depersonalized them, impersonalized them, and hidden their mystical powers in the innocent quantities of physics. But physics needs no powers. *Agents* – we – need powers (to *be* agents), and we have powers. That is what decision theory is *for* – to help us employ our powers wisely.

Maybe in some remote metaphysical sense this is a fraud – maybe there *are* no agents. But this is surely not something for which we have any evidence at all, while, on the contrary, we have a lot of relatively direct evidence that we *are* agents: not only the anguish of indecision, but the development of sophisticated decision theories.

The upshot is that if we construe causality in terms of the powers of inanimate objects, we are indulging in superstition. These powers are the powers of agents. To attribute them to sticks and stones (or subatomic particles) is a primitive fairy tale.

Note that we are *not* identifying causality with uniformity. Causality is associated with uniformities – but not all uniformities. It is associated with the uniformities through which agents, actual or hypothetical, can exercise their power over the world. *Can*, in principle. We could change the period of the earth by changing the mass of the sun, but not vice versa.

3

So what about probabilistic causality? Let's suppose the previous analysis of deterministic causality is on the right track. Physics itself has no use for causal powers, but its laws do characterize uniformities that we, as agents, can often use to exercise our power to make things occur in the world. Statistical laws have even less use for causal powers. But they can function in exactly the same way. Suppose you have a red die and a green die. The red die is so loaded that it always gives a one; the green one is so loaded that it always gives a six. (The dice are characterized in terms of uniformities.) I can cause a six or a one by choosing to roll the appropriate die. Suppose that the dice are only a bit loaded. I can still control matters somewhat by choosing the reference class of which the next roll is to be a member: rolls of the red die or rolls of the green die.

But aren't we talking about chances here? Isn't it that the chance of a six on the red die differs from the chance of a six on the green die?

Chances, like causes, are associated with agents. The original red die did not *cause* a six to appear; *I* caused the six to appear by rolling the red die. Similarly in the chance case: I "increased the chances" of a six by choosing to roll the red die. That is, I chose the reference class of the next roll to be the set of rolls of the red die.

Now I must grant that finding the right reference class is not always trivial. If (for example) the red die in question is to be tossed once and destroyed, tosses of *that* die can't be the right reference class. But tosses of dice with the same density distribution might do, or maybe physical mechanisms subjected to any of a class of initial stimuli. . . . We can not only find a class large enough to support an appropriate relative frequency, but one such that our belief in that relative frequency is well founded (perhaps on theoretical grounds rather than on direct statistical grounds).

In the past I have expressed a certain limited sympathy with the desire to deal with chances rather than frequencies. For example, in considering certain ideal objects (e.g. a dodecahedron) that may never exist, we may consider what their behavior would be like in the arbitrarily long run. Or an ideal pointer spinning on an ideal disk. We may take chances, in such cases, to be the limits (if any) to which we have good reason (if any) to think that actual frequencies (if any) would approach. A more tantalizing use of the notion occurs in quantum mechanics; it may well be that it is simply an implication of the theory that a certain preparation gives a certain distribution of chances over outcomes. In that case I think "chance" is best construed as a primitive term of that special theory, having (almost) nothing to do with ordinary agent-oriented talk of chances.

Under ordinary circumstances, and in particular in decision theoretic situations, I think quantum type chance has absolutely nothing to do with our decision procedures. Just as it is an absurdity of most subjective interpretations of probability that it is supposed that there is a "degree of belief" that an agent has in each proposition of his language, that can be represented by a unique real number, so, I suggest, it is an equally perverse absurdity to suppose that there is a real-valued "chance" in any situation that a given outcome will ensue. It may be that there is some quantum mechanical "chance" that if I smoke I will get cancer, but for me to suppose that there is utterly gratuitous and unnecessary for my deliberations.

Leaving aside the rarefied realm of quantum mechanics, we have no need for "objective chances"; nor, of course, do we need anything like subjective degrees of belief.

Just to make things really sporting, let us do without any sort of conditional connective in the object language, too. All counterfactuals and hypotheticals we will construe as syntactical: that is we represent them by means of hypothetical changes in a set of sentences representing a corpus of knowledge. As has been pointed out by many, there are difficulties with this. I'm just assuming that the difficulties are not insuperable – at least within the decision-theoretic context.

Obviously, we cannot do altogether without modal locutions. We must be able to say that something is within an agent's power, and that is already modal. But we don't need anything in the object language – no conditional connectives – and we don't have to qualify modal statements with probabilities.

Having thrown away all our wonderful modern weapons, can we put up a good fight against the monster of decision theory? Can we recapture, in appropriate ways, the truths captured by causal and epistemic analyses, where those analyses seem to work, and avoid the difficulties of those theories where those theories seem not to work? Let's see.

4

One thing we will give ourselves – to my mind much less superstitious than causal connections in the world, or lurking propensities or chances or other ghosts – is the phenomenal reality of choice. If people don't make decisions, decision theory is either a fraud or a bore.

A few preliminaries are in order. The interpretation of probability we use is epistemic; it is relativized to a body of knowledge, but it is *objective* in the sense that, given a body of knowledge, all probabilities relative to that body of knowledge are determined on logical grounds alone. Probabilities are *based on* frequencies (or, perhaps in some *theoretical* cases, chances – these cases will not concern us here). If the probability that the next roll of this die yields a six is the interval $[p,q]$, then there is *some* reference set – it need not be rolls of this die, nor even have anything to do with dice – in which *some* property (appropriately related to this roll yielding a six) is *known*, in the body of knowledge relative to which we are judging, to fall in exactly the interval $[p,q]$.

What makes the connection between the known interval in a reference class and the corresponding probability, is the relation of randomness. Choosing the right reference class – that is, the reference class of which an object is to be construed as a random member – is tricky, and I won't go into it here [2] except in one respect. Among the rules for getting the appropriate reference class (on my view) is the *strength* rule. This rule states (roughly) that if there is no disagreement between a reference class yielding an interval I and any other class, and one between a reference class yielding an interval I' and any other class, and I is a properly included in I', then we should take as reference class the one yielding I rather than the one yielding I'. In short, we should use the most informative reference class we can get away with.

Issac Levi has shown that it is exactly this strength rule that leads to conflict between my epistemic probability and the principle of confirmational conditionalization. On this matter, completely unpersuaded by all forms of Dutch book arguments, I bite the bullet. So much the worse for confirmational conditionalization. The matter becomes important also in connection with decision theory.

Conditional probability is important to us, of course, but we do not define it in terms of unconditional probabilities. On the contrary, we define the conditional probability of A given B (relative to a body of knowledge K) to be the *unconditional* probability of A relative to the logical consequences of K and the unit class of B:

$$Prob(A,B,K) = Prob(A,L\text{-}Cn(K \cup \{B\})).$$

We can define expected utility in a perfectly conventional way with this kind of probability. If the value of a certain outcome is U, and its probability is $[p,q]$, its utility is the interval $[Up,Uq]$. I believe, though I do not know first hand, that some economists have considered measuring values by intervals – this would introduce no essentially new complications. If the value of the outcome is $[U,V]$, the expected utility is $[Up,Vq]$.

In order to deal with the expected utility of an act that may have any of several outcomes, we must be able to deal with probability *distributions*. Let $S_0,S_1 \ldots S_n$ be a set of sentences, and suppose that relative to K, the probability of each of these sentences is determined by what we know of a distribution of some random quantity in the reference set R. That is, we know of some object a that it belongs to R, and, relative to what we know, it is a *random member* (in the epistemic sense) of R with respect to having any particular value of Q. Suppose, finally, that what our body of knowledge K contains by way of knowledge of the distribution of Q in R is that its distribution is one of a closed convex set of distributions P.

Suppose that $A = \langle A_0, A_1, \ldots, A_n \rangle$ is the set of types of acts contemplated, and that the specific, dated, act is a. The alternatives may be expressed as "$a \in A_1$", "$a \in A_2$", etc. These will be referred to as "the acts". Similarly let $O = \langle O_0, O_1, \ldots \rangle$ be the set of types of outcomes we consider seriously possible – shades of Levi! – and "$a \in O_0$", "$a \in O_1$", ... the specific outcomes we wish to take account of in the present situation. These we call "the relevant outcomes". Suppose that for each act the probabilities of the relevant outcomes are determined by the same reference class, say $R(A)$. Let P be the closed convex set of distributions known in K to characterize $R(A)$. The (interval) utility of an act "$a \in A_i$" is then defined as

$$U(\text{“}a \in A_i\text{”}) = \left[\min_{p \in P} U(\text{“}a \in O_j\text{”}) * p(\text{“}a \in O_j\text{”}, \text{“}a \in A_i\text{”}, K),\right.$$
$$\left.\max_{p \in P} U(\text{“}a \in O_j\text{”}) * p(\text{“}a \in O_j\text{”}, \text{“}a \in A_i\text{”}, K)\right]$$

The principle of dominance carries over unchanged into this framework. Let us say that A_1 dominates A_2 just in case the lower expected utility of A_1 is greater than the upper expected utility of A_2: There is no circumstance that we consider physically possible in which the expected utility of A_2 can exceed the expected utility of A_1. An act that is dominated is quite clearly inappropriate relative to what we are taking ourselves to know.

Naturally, there may be any number of undominated acts, characterized by distinct utility intervals. How one chooses among the undominated acts is open to discussion. One might maintain that there simply are no further principles of rational choice: once one has ruled out the dominated acts, one has exhausted the resources of rationality, and must turn to morality proper, aesthetics, tea leaves, minimax, etc. (Note that stochastic devices – flipping coins – are no help; a stochastic mixture of two undominated acts is itself undominated and does not dominate either of the two undominated components.)

All this we may leave to one side for now. What concerns us is the role of causal considerations in decision theory. And in particular the difficult and anomolous situations so much discussed in the decision-theoretic literature. Given the machinery we have outlined, and given a profound distaste for metaphysical obscurantism, what can we say about these curious situations?

What I propose to do is to bite the existential bullet by developing the approach to these tricky decision problems that I first suggested in *Theory and Decision* [3]. The idea is this: to the extent that I am actually making a choice, I must regard that choice as free. To regard it as free is exactly to regard it as *without* the evidential relevance it would have if it were regarded, not as an *act*, but as a bit of behavior. I cannot construe my act as evidence without depriving it of the character of an *act*. That is, we must maintain a sharp distinction between an act, and a specimen of behavior. The same physical sequence of events, from your point of view, may be an act, while from my point of view it is a mere bit of behavior. The way we evaluate all this evidentially is in terms of the choice of a reference class. I may, of course, look back on an act of mine as a bit of behavior – that is what my psychoanalyst tries to do, right? – and thus construe it as evidence that changes the reference class. But I can't do this simultaneously with *making* the choice. When I make the choice, I must construe the reference class as wide enough to admit both possibilities.

For useful examples, let us look at some of Gibbard and Harper's interesting stories. King David has a choice of two actions, sending for Bathsheba or abstaining; he can make $d \in B$ true or he can make $d \in nB$ true. This choice is within his power. Since this (in the story) is construed as justice or lack thereof, it is in this power to make $d \in J$ or $d \in nJ$ true. Let K be the general potential reference class of reigns of kings, and JK and nJK be those of just and unjust kings. David can make himself a member of either of those classes, and in fact, by his act d, a random member of either one, with respect to being revolted against, relative to what he knows.

Construe d from outside as a bit of behavior. The same analysis goes through. If you or I see David sending for Bathsheba, and know just what he knows, his reign becomes a random member of the reigns of unjust kings, and the probability of his being revolted against changes accordingly.

The case of Solomon is different. Solomon knows, and we suppose ourselves to know, that it is kings without charisma who are successfully revolted against. Charisma is not within the agent's power to change. But there is a correlation between charisma and being just. Does sending for Bathsheba constitute evidence that Solomon lacks charisma, and therefore does it increase the likelihood that he will be successfully revolted against? For us, yes; for him, no. But why?

Because he has no power over whether or not he has charisma, and therefore no power over whether or not he will be revolted against. He does have power over whether or not be performs the unjust act *B* of sending for Bathsheba. But since the act is under his control, it has no evidential value for him. If he was a random member of the set of Kings, with respect to being revolted against, prior to his choice, he remains one after his choice. A free act has no evidential value for the agent.

But this is not to say that it does not have evidential value for us. *We* can see it as a mere piece of behavior, and as such it has evidential value concerning Solomon's charisma or lack of it. In retrospect, when it is no longer an act that he is performing, but a piece of behavior he has exhibited, Solomon can see it that way, too.

Finally, let us look at Gibbard and Harper's example of the executive up for promotion. You will recall that he has already taken all his examinations, and knows that he is tied with other candidates for the promotion in all aspects but one, ruthlessness. Further, he and all the other candidates have taken a ruthlessness test, and been graded on it, and know only that the candidate with the highest score has been given the promotion. A situation arises in which he can display ruthlessness. If he does so, it is evidence that he has done well on the ruthlessness test. That, in turn, is evidence that he has won the coveted promotion. Harper and Gibbard correctly observe that this would be absurd: the executive cannot raise the probability of his getting the promotion at this point in time, because there can be no causal connection, i.e. he has no *power* over his prospects, between his action now and his prospects for promotion.

On the other hand, it is clear (though not stressed by Gibbard and Harper) that if *we* see the executive being ruthless, that will increase the probability, for *us*, that he is the one who got the promotion. We will be in a position to offer higher odds in betting that he got his promotion. We become rationally more confident that he got the promotion. There is still no causal connection. Furthermore, more astonishing yet, might not the executive be in the same position as we? He may have acted automatically, but if we can get him to reflect (uncharacteristically) on his display of ruthlessness (or lack of it) that reflection should be exactly as compelling rationally for him as for us – it should change his degree of confidence that he was given the promotion.

But obviously, we don't want the executive to do a mean thing just to increase his subjective confidence that he is mean enough to have got the job.

The same analysis works. For the executive, as for us, before the opportunity to be ruthless, he is a random member of some class *R* with respect to being of ruthless character. The probability for him, as for us, that he won the ruthlessness contest is (say) $[p,q]$, because that is the proportion of members of *R* that are ruthless enough to win.

Now the opportunity strikes. The executive has a chance to be ruthless. But alas, he has no (short term) power over his character, and that is what the test tests. So even though he has the power to be mean to his subordinate, the exercise of that power will not put him in a new reference class for determining the probability that he won the meanness contest. The probability will remain unchanged that he will be promoted.

From *our* point of view, his choice *is* significant; it provides *us* with evidence concerning his character, and therefore concerning whether or not he won the promotion. Similarly for the executive himself, when he looks *back* on what he has done as a piece of behavior, that piece of behavior can function as evidence. But while he is deliberating and acting, his actions and deliberations must continue to allow for all the alternatives; they cannot place him in any significant subsets for the calculation of probabilities. We must distinguish, then, between contemplating *acting* in a certain way, and contemplating *having acted* in that way.

There are a lot of details to be worked out before this becomes a viable foundation for decision theory. But I think it captures an important aspect of decision making – even though it is an aspect that gets us involved with the theory of action – and is worth pursuing.

University of Rochester

REFERENCES

1. *Theory and Measurement*, Cambridge University Press (1984).
2. 'The reference class', *Philosophy of Science* **50**, 374–397 (1983).
3. 'Acts and conditional probabilities', *Theory and Decision* **12**, 149–171 (1980).

PART II

RATIONAL BELIEF CHANGE

PETER GÄRDENFORS

CAUSATION AND THE DYNAMICS OF BELIEF

1. THE INFORMATION PROBLEM

The basic idea of probabilistic causality, as introduced by Reichenbach (1956), Good (1961–62), Suppes (1970), is that a cause raises the probability of the effect. However, not all events that raise the probability of another event can be counted as causes. In order to sort out those pairs of events that stand in a causal relation to each other, further conditions must be added. Suppes handles this by distinguishing between genuine and spurious causes as well as between direct and indirect causes.

However, it has been claimed by several authors that Suppes' account of causation is defective (see e.g. Hesslow, 1976; Salmon, 1980; and Otte, 1981) and a number of troublesome examples have been presented. Some of these examples will be discussed in ths sequel. In my opinion, the common feature of these examples is that they show that the *information* contained in a single probability function is not sufficient to differentiate genuine from spurious causes or direct from indirect causes. Otte (1981) summarizes his analysis of Suppes' definition as follows:

> The basic problem facing a definition of spurious causation which depends solely on probability relations among events seems to be that there is no way of distinguishing between a causal chain and a fork when the probabilities involved are the same. Suppose we have the chain of necessary and sufficient causes $C \rightarrow B \rightarrow A$ and a necessary and sufficient common cause $B \leftarrow C \rightarrow A$. All of the probability relations in these two cases will be identical, yet in one case B is a genuine cause of A and in another case B is a spurious cause of A. Thus there appears to be no way to distinguish genuine from spurious causes using only probability relations among the events, which is a very serious problem for a probabilistic theory of causality. (p. 180)

Thus the problem for the theory seems to be a problem of insufficient information: a single probability function does not contain enough information to enable us to sort out genuine causal relations from spurious ones. I now turn to a presentation of some ways of supplementing a probability function with further information.

William L. Harper and Brian Skyrms (eds.), Causation in Decision, Belief Change, and Statistics, II, 85–104.

Suppes himself mentions a related problem in his discussion of the definition of "spurious cause in sense one" (Suppes, 1970, pp. 23–26). His solution to the problem, i.e. the definition of "spurious cause in sense two" consists in introducing a partition of the event space in relation to which the causal relations are evaluated. This is one way of taking further information into account when defining causal relations. Suppes' idea of using partitions has been elaborated by Cartwright (1979) and Skyrms (1980) among others. The partitions they use are supposed to be maximal specifications of causally relevant background factors. I believe that something like this was also the intended interpretation of the partitions used by Suppes.

One drawback, in my opinion, of Cartwright's and Skyrms' proposals is that they are partially circular: in order to distinguish between genuine and spurious causes we must know what are the causally relevant background factors. So even if they succeed in avoiding some of the counterexamples that have been presented for Suppes' theory, their definitions are not complete analyses of probabilistic causation. Furthermore, the extra amount of information demanded is quite substantial: in order to determine whether C is a cause of E, all causally relevant background factors must be available. It seems clear that we often make claims about causal relations between events, even if we do not know exactly which factors are causally relevant to the events in question.

Another way of exploiting additional information and avoiding the alleged counterexamples is to say that if we only knew more about the process between a cause C and an effect E, we would see that each step in the chain of events will satisfy Suppes' definitions, even if the direct step from C to E does not. Or, in other words, if we insert a sufficient number of steps between C and E, the counterexamples will disappear. For example, Rosen (1978) writes in her comment on Hesslow's (1976) thrombosis example (to be discussed in Section 4):

> We wrongly believe that taking the pills always lowers a person's probability of thrombosis because we base our beliefs on an inadequate and superficial knowledge of the causal structures in this medical domain where unanticipated and unappreciated neurophysiological features are not given sufficient attention or adequate weighting. (p. 606)

Salmon (1980) calls this strategy 'the method of more detailed specification' (p. 64). It is clear that if this strategy is to be applied to any causal process, we must assume more or less total information about the process in order to evaluate causal claims. Thus this strategy demands even more additional information than the strategy of using partitions. From a pragmatic point of view, another counterargument against this strategy is that we frequently make causal claims even if we only have a limited knowledge of the intermediate steps or the underlying processes.

2. AN EPISTEMIC APPROACH TO CAUSALITY

My aim is to give an epistemological analysis of some aspects of causality. I want to say something about what kind of beliefs our causal claims are based on. Probabilities are essential, but the only probabilities I will consider are subjective probabilities.

I do not know if it is possible to give a 'realistic' or 'objective' interpretation of causality where the causal relation holds among real events independently of cognitive processes, but such an interpretation will not be discussed here. For me, causal claims are interpreted *in relation to a given state of belief*. The basic relation to be analysed is "in the state of belief P, C is a cause of E", where C and E are beliefs about the occurrence of two single events. The central problem is to delimit the type of information that is used when making a causal claim of this type. Only in the concluding section will I comment on causal relations between general events.

My analysis is a combination of two seemingly disparate ideas. The first is the central doctrine in Bayesian epistemology that beliefs can be represented by probabilities. I will take this thesis as unproblematic in the present context at least for beliefs concerning single events.

The second idea is based on Hume's second definition of causality, i.e. if the cause had not been, the effect never had existed. The status of this definition will be discussed in the final section. David Lewis (1973) took this defiition as the starting-point for his counterfactual analysis of causation, using possible worlds and similarities between possible worlds as the primitive concepts. Instead I will give an epistemological interpretation of Hume's definition. The central notions are probability functions and what I call 'contractions' of probability functions. A contraction of a state of belief consists in giving up, i.e. no longer accepting, some of the beliefs that were accepted in the earlier state of belief. Here I treat a proposition C as accepted in the state of belief represented by P, iff $P(C) = 1$. Since states of belief are assumed to be represented by probability functions, a central problem is to give a characterization of the contraction of a probability function. This will be the objective of Section 3. Here I only introduce the notation $P-C$ for the contraction of P with respect to C, where C is the proposition to be retracted.

Assuming the concept of contraction as already given, I can now state my version of Hume's second definition of the causal relation:

DEFINITION 1. In the state of belief represented by P, C is a (direct) cause of E, if and only if (i) $P(C) = P(E) = 1$ and (ii) $P-C(E/C) > P-C(E)$.

This definition is a probabilistic version of Hume's second definition. The main point of the definition is to put the contraction $P-C$ at the focus of the evaluation of a causal claim instead of the probability function P. In P both C and E have maximal probability, according to the first clause of the definition, so these values cannot be used for any interesting comparisons. The probability function $P-C$ represents the epistemic situation that results from P if the belief that C has occurred is given up, and it is relative to this situation that the epistemic status of E is evaluated. The evaluation is simply the key idea in Suppes' theory: the cause raises the probability of the effect. The difference lies in which probability function this test is applied to.

Let us illustrate Definition 1 by a simple example. Assume that C describes the event that the butcher consumed half a pound of salt and that E describes the event that he died the night after. Let P be the belief function corresponding to the present state of belief, in which it is known that C and E have occurred, i.e. $P(C) = 1$ and $P(E) = 1$. If we want to determine whether C caused E, we should, according to Definition 1, consider the belief function $P-C$, corresponding to the epistemic state where the belief in C has been retracted, but which in other respects is as similar to P as possible. This is the hypothetical state where we assume that we did not know that the butcher ate the salt. Since salt is a dangerous poison if consumed in large quantities, the probability of the butcher dying, given that he eats half a pound of salt, increases drastically, i.e. $P-C(E/C) > P-C(E)$, and thus C is a cause of E according to Definition 1.

It should already be noted here that, unlike Suppes, I do not present a definition of a *prima facie* cause, which must then be complemented by a distinction between spurious and genuine causes. In Section 4, I will try to show that Definition 1 can handle both the cases of spurious causes discussed by Suppes, as well as the putative counterexamples that his theory has been charged with. Also note that the definition is intended to cover the deterministic case as well, i.e. the case when all relevant probabilities are 1 or 0.

The additional information assumed in Definition 1, apart from the probability function P representing the given state of belief, is a probability function $P-C$ for each event C. The presentation of contraction functions in Section 3 will better show how much additional information this assumption amounts to. I think that it will then be clear that, from an epistemological point of view, this additional information is more naturally available than both maximal specifications of causally relevant background factors and total information about the course of events as discussed earlier.

3. CONTRACTIONS OF PROBABILITY FUNCTIONS

A central part of Bayesianism is that rational changes of belief can be represented by conditionalization whenever the information to be added is consistent with the given state of belief as represented by some probability function. However, there are other kinds of changes of belief that cannot easily be modelled by the conditionalization process. Sometimes we have to revise our beliefs in the light of some evidence that contradicts what we had earlier mistakenly accepted. And when $P(E) = 0$, where P represents the present state of belief and E is the new evidence to be accommodated, the conditionalization process is undefined.

And sometimes we give up some of our beliefs. This kind of change of belief will be called a contraction and the goal of this section is to present a way of modelling this process. One kind of situation where contractions occur is in debates. Here the opponents hypothetically give up their beliefs in the point at issue in order to avoid begging the question. Another type of situation is explanations, where belief in the event to be explained is given up. A theory of explanations based on contractions is presented in Gärdenfors (1980). And a main thesis of this article, as should be clear from Definition 1, is that contractions are central for the evaluation of causal claims.

When contracting a state of belief with respect to a belief C, it will be necessary to change the probability values of other beliefs as well in order to comply with the axioms of probability calculus. However, there are in general several ways of fulfilling these axioms. For example, I am certain that all presidents of the U.S. are or have been citizens of the U.S., that Ronald Reagan is the present president of the U.S., and, consequently, that Reagan is a citizen of the U.S. Suppose that I now, for some reason, contract my set of beliefs by giving up the belief C that Reagan is a citizen of the U.S., i.e. I change my probability function P where $P(C) = 1$ to a function $P - C$ where $P - C(C) < 1$. I will then have to retract either my belief that all presidents of the U.S. are citizens of the U.S. or my belief that Reagan is the president of the U.S. As long as only probability axioms are concerned, both ways are open.

An important problem concerning contractions is how one determines which among the accepted beliefs, i.e. those C's where $P(C) = 1$, are to be retained and which are to be removed. One requirement for contraction that can be formulated in relation to the example given above is that the two beliefs should not both be retracted since the loss of information should be kept as small as possible so that no belief is given up unnecessarily.

In Gärdenfors (1986), I have formulated a number of postulates for contractions of probability functions. These postulates are based on the idea that the contraction $P - C$ of a probability function P with respect to C should be as small as possible in order to minimize the number of beliefs that are retracted. In a sense that will be made more precise later, contractions can be viewed as 'backwards' conditionalization. The postulates for probabilistic contractions can also be regarded as generalizations of a set of postulates for contractions in the deterministic case which I and others have developed elsewhere (cf. e.g. Harper, 1978; Gärdenfors, 1979, 1981, 1982; Alchourrón, Gärdenfors and Makinson, 1985).

I will here give a brief presentation of the postulates for probabilistic contraction. Formally, this process can be represented as a function from $P\# \times E\#$ to $P\#$, where $P\#$ is the set of all probability functions and $E\#$ is the space of events over which these functions are defined. (I will here assume that the events in $E\#$ are described in some language which includes the standard propositional operators and which is ruled by a logic including classical propositional logic.) The value of such a contraction function, when applied to arguments P and C, will be called the contraction of P with respect to C and it will be denoted $P-C$. Let us say that an event C is accepted in the state of belief represented by P iff $P(C) = 1$.

The first postulate is a requirement of 'success' simply requiring that C not be accepted in $P-C$, unless C is logically valid (i.e. describes the unit event), in which case it can never be retracted:

(−1) $P-C(C) < 1$ iff C is not logically valid.

It should be noted that this postulate does not say anything about the magnitude of $P-C(C)$. This leaves open a range of possibilities for an explicit construction of a contraction function. None of these possibilities will be ruled out by the remaining postulates. The value of $P-C(C)$ can be seen as a parameter in the construction of $P-C$.

The second postulate requires that the contraction $P-C$ is only dependent on the content of C, not on its linguistic formulation:

(−2) If C and D are logically equivalent, i.e. describe the same event, then $P-C = P-D$.

The following postulate is only needed to cover the trivial case when C is already not accepted in P:

(−3) If $P(C) < 1$, then $P-C = P$.

So far, the postulates have only stated some mild regularity conditions. The next one is more interesting:

(−4) If $P(C) = 1$, then $P-C(D/C) = P(D)$, for all D in $E\#$.

This means that if C is first retracted from P and then added again (via conditionalization), then one is back in the original state of belief. This postulate, which will be called the recovery postulate, is one way of formulating the idea that the contraction of P with respect to C should be minimal in the sense that unnecessary losses of information should be avoided. It also makes precise the sense in which contraction is 'backwards' conditionalization.

The postulate (−4) is the probabilistic version of a corresponding postulate in the deterministic case. If this postulate is accepted, then Definition 1 can be simplified. According to clause (i) the cause C and the effect E are already accepted as having occurred, i.e. $P(C) = P(E) = 1$. If we apply (−4), we know that $P-C(E/C) = P(E) = 1$, so this means that Definition 1 reduces to the requirement that $P-C(E) < 1$.

This simple requirement is perhaps one way of formulating Hume's second definition "if the first object (i.e. C) had not been, the second (i.e. E) never had existed". However, in the typical case of a probabilistic cause, the postulate (-4) is too strong. To give an example which shows the problems of (-4), assume that we have thrown a die (event C) and that it showed a six (event D). In the probability function P representing the present state of belief C and D are accepted as having occurred, i.e. $P(C) = P(D) = 1$. The contraction $P-C$ is most naturally taken to be identical with the state of belief immediately before the throw. (In this state $P-C(C)$ may be high, but it is not 1 until the throw has actually been made.) However, $P-C(D/C)$ is not 1 as required in (-4), but only 1/6. On the other hand, the throw C makes D an element of a homogeneous partition.

In general it is thus more natural to require something like the following weaker version:

($-4'$) In $P-C$, C makes D an element of a homogeneous partition.

(-4) is then the special case of ($-4'$) when D is the only element of the partition.

The final postulate is more complicated and concerns the connection between $P-C$ and $P-C\&D$:

(-5) If $P-C\&D(-C) > 0$, then $P-C(E/-C) =$
$P-C\&D(E/-C)$, for all E.

In order to understand this postulate, we first present one of the arguments that has been proposed as a justification for conditionalization. Unlike all other changes of P to make C certain, conditionalization does not distort the probability ratios, equalities, and inequalities among sentences that imply C (cf. Teller, 1976; and Lewis, 1976, p. 311). In other words, the probability proportions among sentences that imply C are the same before and after conditionalization.

Now, if contraction may be regarded as 'backwards' conditionalization, then a similar argument should be applicable to this process as well. More precisely: when contracting P with respect to C, some sentences that imply $-C$ will receive non-zero probabilities and when contracting P with respect to $C\&D$ some sentences that imply $-C$ or some sentences that imply $-D$ (or both) will receive non-zero probabilities. If, in the latter case, some sentences that imply $-C$ receive non-zero probabilities, i.e. if $P-C\&D(-C) > 0$, then the two contractions should give the same proportions of probabilities to the sentences implying $-C$, i.e., $P-C(E/-C)$ should be equal to $P-C\&D(E/-C)$, for all E. But this is exactly the content of (-5).

This completes the set of postulates for probabilistic contraction functions. It should be noted that the postulates do not determine a unique contraction function, but they only introduce rationality constraints on such functions. Among other things the value of $P-C(C)$ can be any number greater than 0 and smaller than 1. It is argued in Gärdenfors (1984) that rationality constraints are not enough to determine a unique contraction function (just as the probability axioms do not determine a unique rational probability function), but that pragmatic factors must be added in order to single out the actual contraction function. In the article, I introduced an ordering of 'epistemic importance' among the beliefs, to be used when determining which beliefs are to be given up when forming a particular contraction. The heuristic rule is that when we have to give up some of our beliefs, we try to retain those with the greatest epistemic importance.

I will conclude my discussion of probabilistic contraction by showing how it relates to *revisions* of states of belief. A revision of a state of belief consists in accepting a sentence, the negation of which was accepted earlier. This kind of change may occur when we receive compelling evidence that contradicts our present beliefs. Sometimes, such a change is also made for the sake of an argument. Revisions in this sense cannot be modelled by standard conditionalization since if C is the contradictory evidence to be added, we have, by definition, $P(C) = 0$, where P is the present subjective probability function.

As before, the revision process can be represented by a function from $P\# \times E\#$ to $P\#$. The revision of a function P with respect to an event description C will be denoted $P{*}C$. In Gärdenfors (1986) a number of postulates are formulated for this process. It can be shown that they are essentially equivalent to the postulates for so called Popper functions (as presented by Popper, 1959; Stalnaker, 1970; Harper, 1976; and van Fraassen, 1976).

It has been argued (by e.g. Levi, 1977; and Gärdenfors, 1981, 1984) that a revision with respect to C can be analyzed as the composition of a contraction with respect to $-C$ and an addition of C to the contracted state of belief. In the terminology of probability functions, this idea can be formulated as follows: in order to arrive at the revision $P{*}C$ one first contracts P with respect to $-C$ and then conditionalizes on

$P-(-C)$ by C. This leads to the following definition of a revision function in terms of a contraction function:

DEFINITION 2. For all probability functions P and all sentences C, $P{*}C(D) = P-(-C)(D/C)$, for all D.

In Gärdenfors (1986) it is established that the revision function defined in this way satisfies all the proposed postulates and thus it is essentially a Popper function.

It is also possible to reverse this construction and define contractions in terms of revisions. The idea is that the beliefs in a contraction $P-C$ can be described as a mixture of what is believed in P and what would be believed in the revision of P with respect to $-C$. For a precise formulation we need the concept of a mixture of two probability functions P and P': (PaP') is the a-mixture of P and P' defined by the equation $(PaP')(C) = a \cdot P(C) + (1-a) \cdot P'(C)$, for all C, where $0 < a < 1$.

DEFINITION 3. For all probability functions P and all sentences C, $P-C(D) = (PaP{*}(-C))(D)$, for all D, where $1 > a = P-C(C)$ iff $P(C) = 1$ and $a = 1$ otherwise.

In other words, $P-C$ is a 'compromise' between the states of belief represented by P and $P{*}(-C)$, where 'a' is a measure of the degree of closeness to the beliefs in P. Note that, for a given revision function, Definition 3 yields different contraction functions depending on the choice of 'a', i.e., the value of $P-C(C)$.

It is shown in Gärdenfors (1986) that if the revision function satisfies the proposed postulates, i.e. if it is a Popper function, then any contraction function introduced via Definition 3 will satisfy the postulates (−1)–(−5). This result and those previously mentioned show that there is an intimate connection between contractions and revisions of probability functions and that either process can be taken as a primitive one and the other then be introduced by Definition 2 or Definition 3.

4. ANALYSIS OF SOME EXAMPLES

After this semiformal presentation of the rational rules that govern contractions of probability functions, it is now time to show the power of the test for causality introduced in Definition 1 by applying it to some of the examples that have occurred in the literature on probabilistic causation. In these examples, the contractions of states of belief that will be used will only be presented in an informal 'intuitive' way and I do not intend to define explicit contraction functions. However, even if this non-technical strategy is used, the postulates for contractions presented in the previous section will serve as regulative ideals.

The first thing that must be shown is that Definition 1 in itself can be used to distinguish between genuine and spurious causes. Let us consider the classical barometer example. Here three events are involved: the falling barometer reading (F), bad weather (W) and low air pressure (L). In order to test whether F causes W we must apply Definition 1 to a state of belief, represented by P, where F, W and L are known to have occurred. The crucial step is the evaluation of the contraction $P-F$. Now, if we give up the belief that F has occurred, the requirement that we should change P as little as possible would, in the most natural interpretation of the example, then make us say something like "if the barometer had not been falling, it must have been malfunctioning or disturbed in some way, since pressure was low". This means that we would keep the belief in L in $P-F$ and thus also the belief in W. In other terms, $P-F(W) = 1$, so $P-F(W/F)$ cannot be greater than $P-F(W)$. Definition 1 thus yields the result that F is not a cause of W as desired. This line of reasoning can, of course, also be expressed by saying that L screens off F from W. However, as will been seen from the following examples, the analysis in terms of contractions can also be applied to cases where the screening off test does not apply.

On the other hand, L is a cause of W (and of F) according to Definition 1. If we give up our belief in L, it is difficult to retain the belief in W, so $P-L(W)$ will be lower than 1 and most naturally of the same magnitude as $P-L(L)$. And, of course, $P-L(W/L) > P-L(W)$, as desired. The reason why the falling barometer F has been called a spurious cause of W is that also F will raise the probability of W in $P-L$, since it is an indication of L. But the fact that $P-L(W/F) > P-L(W)$ is not sufficient to make F a cause of W according to Definition 1. I believe that the strategy of this example can be used to analyse spurious causes in general.

The second example I want to discuss is due to Germund Hesslow (1976). He suggests (p. 291) that the example shows that it is possible that a cause can lower the probability of the effect. The example starts from the claim that taking contraceptive pills can cause thrombosis. But pregnancy can also cause thrombosis, and taking the pills lowers the probability of pregnancy. It is perfectly possible that pregnancy is a stronger cause of thrombosis than contraceptive pills, so that taking contraceptive pills may lower the probability of thrombosis. Thus this example seems to show that the basic thesis of probabilistic causation, i.e. that the cause should raise the probability of the effect, is violated.

When analysing this example in relation to Definition 1, we should start from a particular situation where a woman the been taking contraceptive pills (C) and suffers a thrombosis (T) (and where she is not pregnant ($-B$)). If we want to determine whether C caused T we should, according to Definition 1, ask what would have happened if she had not been taking the pills. It is a part of the normal context of this example that the woman has been taking the pills because she wants to avoid becoming pregnant (A). So, if she had not been taking pills, she would have avoided becoming pregnant by some other means. Thus A is still very likely in $P-C$, i.e. $P-C(A)$ is close to 1, and so $P-C(B)$ is close to 0. So, even if $P-C(T/B)$ may be higher than $P-C(T/C)$, it is still the case that $P-C(T/C) > P-C(T)$, at least if we assume that nothing else but B and C are relevant to T. This is all we need according to Definition 1 in order to say that C caused T.

Other examples that are troublesome for Suppes' theory are those which involve so called *interactive forks*. Salmon (1980) has presented the following instructive example:

> Pool balls lie on the table in such a way that the player can put the 8-ball into one corner pocket at the far end of the table if and almost only if his cue-ball goes into the other far corner pocket. Being a relative novice, the player does not realize that fact; moreover, his skill is such that he has only a 50–50 chance of sinking the 8-ball even if he tries. Let us make the further plausible assumption that, if the two balls drop into the respective pockets, the 8-ball will fall before the cue-ball does. Let the event A be the player attempting that shot, B the dropping of the 8-ball into the corner pocket, and C the dropping of the cue-ball into the other corner pocket. Among all of the various shots the player may attempt, a small proportion will result in the cue-ball landing in that pocket. Thus, $P(C/B) > P(C)$; consequently, the 8-ball falling into one pocket is a prima facie cause of the cue-ball falling into the other pocket. This is as it should be, but we must also be able to classify B as a spurious cause of C. It is not quite clear how this is to be accomplished. The event A, which must surely qualify as a direct cause of both B and C does not screen B off from C, for $P(C/A) = 1/2$ while $P(C/A \cdot B) = 1$. (pp. 65–66)

The task before us is to show that B is not a cause of C according to Definition 1. The crucial step is to describe $P-B$ in sufficient detail.

Here it is important to keep in mind that the contraction is a minimal change so that as few beliefs as possible are given up in $P-B$. There is one event, not considered by Salmon, which I think should be made an explicit part of the story, namely the collision between the cue-ball and the 8-ball. Let us call this event D. Now, if B had not occurred, and we should keep as many beliefs as possible in $P-B$, then the belief in D, and consequently also the belief in C, should be retained. We can, for example, imagine that we do not know whether the course of the 8-ball has been interrupted after the collision, which means that the probability of B, i.e. $P-B(B)$, is less than one, as it should be, but we still want to have $P-B(D) = 1$. By the conditions of the example, we then know that $P-B(C/B) = P-B(C/D) = P-B(C)$, so B is not a cause of C according to Definition 1.

The reason why Salmon can claim that $P(C/B) > P(C)$ is that he is describing another state of belief than the one represented by $P-B$. The epistemic situation he has in mind is the one where we know that the player has attempted the shot, but where we do not yet know whether it will result in the collision D or not. (Note that D describes a collision of the kind that sinks the 8-ball). This is the situation that would be described by $P-D$ in the present framework. And in this situation, the occurrence of B is indeed a strong indication of the occurrence of C, i.e. $P-D(C/B) > P-D(C)$. But, as I hope has been shown by the analysis above, $P-D$ is not the correct epistemic situation to use when testing whether B causes C or not. The upshot of this and the previous examples is that one must be careful to specify which epistemic situation is presupposed when probability comparisons are made.

In pictorial terms, the contraction method applied in the analysis consists in going "backwards" along a chain of (more or less hypothetical) epistemic states that lead up to the event which is seen as the effect of the causal chain. If this chain of epistemic states, represented by probability functions, is reversed so that we go "forwards" by conditionalization, the result will be essentially the method of *successive reconditionalization* presented as "a modest suggestion" (pp. 66–70) by Salmon in his article (1980). The main difference between his method and the one presented here is that Salmon presumes an initial epistemic state as a starting point for the reconditionalizations, while the present method starts from the "final" epistemic state.

It may be objected, both to the analysis of the pool balls example and the analysis of Hesslow's thrombosis example, that I need to resort to artificial events like the collision D in the first example and the desire A to avoid pregnancy in the second. Thus it may seem that my analysis is open to the same criticism as was levelled against the strategy demanding full information about the causal chain of events, the strategy that Salmon (1980) called the method of more detailed specification. However, I do not believe that these additional events require any extra information that is not included in the two examples. On the contrary, both events are parts of the background knowledge that goes along with the stories.

5. A COMPARISON WITH GRANGER CAUSALITY

One of the most influential analyses of causality within the social sciences is that developed by Granger. I believe that his analysis, for example, as presented in this book, can be seen as essentially a special case of the analysis suggested in this paper. In order to substantiate this claim, let me first give a brief summary of Granger's analysis.

Granger prepares his definition of the causal relation by formulating two "basic axioms" for causation:

(G1) Temporal Priority: A causal event will occur before the effect.

(G2) Special Relationship: A causal event will contain information that is not contained in any other event occurring no later than the cause.

The intended application area of Granger's analysis is economic variables. He represents such a variable by a random variable sequence $x(t)$ observed at discrete time intervals $j = 0, 1, 2, \ldots, t$. An event which occurs at time t is identified with the value of such a variable at t. The set of all values of the relevant variables up to time t is denoted $O(t)$. We also let $O(t) - y(t)$ denote the information contained in $O(t)$ minus the sequence $y(t-j)$, $j > 0$. His definition can now be formulated as follows:

DEFINITION 4. The event described by $y(t)$ causes the event $x(t+1)$ iff $P(x(t+1) \in A/O(t)) \neq P(x(t+1) \in A/O(t) - y(t))$, for some set A.

This definition is essentially Definition 1 restriction to the special case when all relevant beliefs can be described by the values of the variables included in $O(t)$ and where the cause occurs immediately before the effect. Granger interprets the probability function P in the definition as an objective probability. The states of belief are thus represented by the different sets of the form $O(t)$. $O(t)$ itself corresponds to the present state of belief, i.e. P in Definition 1, and $O(t) - y(t)$ corresponds to the contraction $P-C$. Note that $O(t) - y(t)$ cannot contain the values $y(t-1)$, $y(t-2)$, ... since these can be strongly correlated with $y(t)$ and thus 'screen off' the causal influence of $y(t)$ on $x(t+1)$.

Granger then makes this definition operational by replacing $O(t)$ with a smaller, more practical information set $I(t)$, which is assumed to contain at least $x(t-j)$, $y(t-j)$, $j > 0$, but may also contain some other measurable variables $z(t-j)$, $j > 0$. If now $P(x(t+1) \in A/I(t)) \neq P(x(t+1) \in A/I(t) - y(t))$, for some set A, then $y(t)$ is said to be a *prima facie* cause of $x(t+1)$ with respect to the information set $I(t)$. The reason for this name is that $y(t)$ may fail to be a cause of $x(t+1)$ in $O(t)$ because some variable that is an actual cause of $x(t+1)$ may have been eliminated from $I(t)$, and if this variable is added to $I(t)$ it would screen off $y(t)$ from $x(t+1)$. This is basically the same problem as the one Suppes (1970) tries to solve by his definition of 'spurious' causes. For a comparison between Suppes' and Granger's theories, see Spohn (1983b). Spohn (1983a) presents an essentially epistemic analysis of causation which is somewhat related to the one presented here.

6. DISCUSSION

The causal relation introduced in Definition 1 should be regarded as a relation of direct causation. It turns out that, given the postulates (−1)−(−5), the relation is not transitive in general. In my opinion, this is as it should be: if C causes D and D causes E, then C may, but need not cause E. The latter case can occur when C causes a part of D but another part of D causes E (cf. Granger, 1988). For example, consider a chemical reaction where a solution of silver chloride is poured into nitric acid which causes silver nitrate to precipitate and the acid to turn into hydrochloric acid. Then one can say that adding the silver chloride is a cause of the formation of the hydrochloric acid, since it contributes the chloride ions. The hydrochloric acid is also a cause of the low pH value of the solution, since it contributes the hydrogen ions. But the silver chloride is not a cause of the low pH value of the solution.

It should be emphasized once again that the analysis of the causal relation given in Definition 1 applies to actual causal chains and not to causal laws in general. The events *C* and *E* that are involved in the test for a causal relation are thus known to have occurred (as is required by clause (i) in Definition 1). This means that the present probabilities of the events are of no use (since they are both 1), but it is other probabilities that must be relied on. Since earlier analyses of probabilistic causation have not been explicit about which epistemic situation these probabilities should be related to it has been easy to slip between probabilities of single events and probabilities of event types as involved in causal laws. Suppes (1970) makes this quite clear:

> A deliberate equivocation in reference between events and kinds of events runs through the earlier systematic sections of this monograph. It is intended that the formalism can be used under either interpretation, . . . (p. 79)

I believe that such an equivocation is dangerous, since which general probability laws are relevant to the evaluation of single events is heavily dependent on the epistemic situation that is presupposed for the evaluation. As I have tried to show in Section 4, many of the problems presented for Suppes' theory and related theories are due to such an equivocation. In my opinion, it is extremely important to use the correct epistemic background situation in the causal analysis.

According to the analysis presented here, the causal relation between single events is the fundamental relation. All general causal claims are derived from causal relations between single events. When we say that stress causes heart attacks, we mean that some events of stress cause some heart attacks. If the relevant background knowledge can be kept relatively fixed, we may even be able to give numerical estimates of how frequently an event of a certain type causes an event of another type. However, such statements are methodologically difficult to evaluate.

On the other hand, when determining the probability of a single event in an epistemic state, actual or hypothetical, probabilistic laws are crucial. It is a fundamental philosophical problem to describe the 'direct inference' from laws to individual events. It is inferences of this kind that will be used when the exact content of the contractions of probability functions is to be determined. Of course, such inferences must be made without relying on the notion of causality in order to avoid circularity. I have nothing to contribute to the solution of this problem, but I want to emphasize that the real substance of the problem of probabilistic causation lies in the probability relations between single events and probabilistic laws. The test for causality formulated in Definition 1 presupposes that the probabilities of the single events are already determined.

Thus a full analysis of causation among the single events consists in two major steps. The first is an account of the relation between probabilistic laws and probabilities of single events. Understanding this relation is, of course, not only important for a theory of causation, but it has applications within many other problem areas. The second step is a test of the probability relations between the single events that are the alleged cause and effect. In this article, Definition 1 has been suggested as an appropriate account of the second step.

This two-step picture of causation can be compared with Hume's two definitions of causation. He first wrote

> ... we may define a cause to be an object followed by another, and where all the objects, similar to the first, are followed by objects similar to the second. (Hume, 1748, Section VII).

Interpreted in modern probabilistic terminology, this definition contains a sketch of a solution to the problem of the connection between the epistemic values of types of events and single events, i.e. the first step in my description. For us, the epistemic values are probabilities while for Hume they are connections between impressions. Hume spends most of his analysis describing the 'genesis' of these connections in our minds.

The second definition follows immediately on the first: "Or, in other words, where, if the first object had not been, the second never had existed." This definition is not supported by any earlier analysis in Hume's text. So Hume's phrase "Or, in other words" seems rather misplaced. When presenting Definition 1, I said that I regard it as a probabilistic version of Hume's second definition. The second definition can thus be seen as a solution to the second step in my description of a causal analysis. If my picture is correct, Hume's two definitions are solutions to two different problems – the first is a 'genetic' analysis of causal connections, and the second is a counterfactual test for when we find a causal relation between two single events.

7. LIMITATIONS OF THE ANALYSIS

The analysis of causation presented in Section 2 is heavily dependent on using probability functions as representations of states of belief. There are, however, several aspects of beliefs that are not captured by such a representation and which effect the outcome of the causal analysis.

One dimension that has been neglected at large in this paper is the time aspect of the beliefs. The description of a single event naturally involves the time it occurred. Some analyses of causation postulate that the cause must not occur later than the effect. If we want this kind of causality it is easy to add the appropriate cause to Definition 1. An alternative approach is not to rule out backwards causation or causal loops *a priori*, but expect that Definition 1, via the properties of the contraction $P-C$, will result in the desired temporal relation between C and E. One way of ensuring this is to postulate that when the probability function P is contracted to $P-C$, the probabilities of all events that occurred before C remain the same in $P-C$ as in P. This means that all beliefs abut the history of events up to C is left unaltered in the construction of the hypothetical state of belief $P-C$.

Another aspect that has been treated rather light-heartedly is the notion of acceptance. As several philosophers have shown, there are many subtleties to acceptance that cannot be captured by simply using probability functions as models of states of belief. Especially Adams' works (e.g. his 1975 book) on probabilities of conditionals are relevant in this context.

Finally, the probabilistic models of belief are not very handy for describing functional dependencies of the kind studied by e.g. Simon and Rescher (1966). As they show, causal relations are often determined by functional correlations between different variables. And such correlations are difficult to describe using only probability functions defined over single events.

In conclusion, I hope to have shown that Definition 1 provides a viable analysis of causality between single events. This analysis reduces the problem of causality, hopefully in a non-circular way, to the problem of identifying contractions of states of belief. In this section, I have indicated some limitations of the analysis which are due to the simplicity of the models of states of belief used in this paper.

ACKNOWLEDGEMENTS

I wish to thank Gerry Hesslow, Isaac Levi and Wolfgang Spohn for helpful discussions.

Department of Philosophy
Lund University
Lund, Sweden

REFERENCES

Adams, E. (1975) *The Logic of Conditionals*, Dordrecht.

Alchourrón, C. E., Gärdenfors, P., and Makinson, D. (1985) 'On the logic of theory change: Partial meet contraction and revision functions', *Journal of Symbolic Logic* **50**, 510–530.

Cartwright, N. (1979) 'Causal laws and effective strategies', *Noûs* **13**, 419–437.

Gärdenfors, P. (1979) 'Conditionals and changes of belief', in *The Logic and Epistemology of Scientific Change*, I. Niiniluoto and R. Tuomela (eds.), Amsterdam.

Gärdenfors, P. (1980) 'A pragmatic approach to explanation', *Philosophy of Science* **47**, 404–423.

Gärdenfors, P. (1981) 'An epistemic approach to conditionals', *American Philosophical Quarterly* **18**, 203–211.

Gärdenfors, P. (1982) 'Rules for rational changes of belief', in *Philosophical Essays Dedicated to Lennart Åqvist on his Fiftieth Birthday*, T. Pauli (ed.), Uppsala.

Gärdenfors, P. (1984) 'Epistemic importance and minimal changes of belief', *Australasian Journal of Philosophy* **62**, 136–157.

Gärdenfors, P. (1986) 'The Dynamics of belief: Contractions and revisions of probability functions', *Topoi* **5**, 24–37.

Good, I. J. (1961–2) 'A causal calculus I–II', *British Journal for the Philosophy of Science* **11**, 305–318, and **12**, 43–51.

Granger, C. W. J. (1988) 'Causality testing in a decision science', *this proceedings*, Vol. I, pp. 3–21.

Harper, W. L. (1976) 'Rational belief change, Popper functions and counterfactuals', in *Foundations of Probability Theory, Statistical Inference, and Statistical Theories of Science*, Vol. I, W. L. Harper and C. H. Hooker (eds.), Dordrecht.

Harper, W. L. (1978) 'Bayesian learning models with revision of evidence', *Philosophia* **7**, 357–367.

Hesslow, G. (1976) 'Two notes on the probabilistic approach to causality', *Philosophy of Science* **43**, 290–292.

Hume, D. (1748) *An Enquiry Concerning Human Understanding*, London.

Levi, I. (1977) 'Subjunctives, dispositions and chances', *Synthese* **34**, 423–455.

Lewis, D. (1973) 'Causation', *Journal of Philosophy* **70**, 556–567.

Lewis, D. (1976) 'Probabilities of conditionals and conditional probabilities', *Philosophical Review* **85**, 297–315.

Otte, R. (1981) 'A critique of Suppes' theory of probabilistic causality', *Synthese* **48**, 167–189.

Popper, K. R. (1959) *The Logic of Scientific Discovery*, London.

Reichenbach, H. (1956) *The Direction of Time*, Berkeley.

Rosen, D. (1978) 'In defense of a probabilistic theory of causality', *Philosophy of Science* **45**, 604–613.

Salmon, W. C. (1980) 'Probabilistic causality', *Pacific Philosophical Quarterly* **61**, 50–74.

Simon, H. A. and Rescher, N. (1966) 'Cause and counterfactual', *Philosophy of Science* **33**, 323–340.

Skyrms, B. (1980) *Causal Necessity*, New Haven.

Spohn, W. (1983a) 'Deterministic and probabilistic reasons and causes', *Erkenntnis* **19**, 371–396.

Spohn, W. (1983b) 'Probabilistic causality: from Hume via Suppes to Granger', in *Causalita e modelli probabilistici*, M. C. Galavotti and G. Gambetta (eds.), Bologna.

Stalnaker, R. (1970) 'Probability and conditionals', *Philosophy of Science* **37**, 64–80.

Suppes, P. (1970) *A Probabilistic Theory of Causality*, Amsterdam.

Teller, P. (1976) 'Conditionalization, observation and change of preference', in *Foundations of Probability Theory, Statistical Inference, and Statistical Theories of Science*, Vol. I, W. L. Harper and C. H. Hooker (eds.), Dordrecht.

van Fraassen, B. C. (1976) 'Representation of conditional probabilities', *Journal of Philosophical Logic* **5**, 417–430.

WOLFGANG SPOHN

ORDINAL CONDITIONAL FUNCTIONS: A DYNAMIC THEORY OF EPISTEMIC STATES*

1. INTRODUCTION

Many of the philosophically most interesting notions are overtly or covertly epistemological. Overtly epistemological notions are, of course, the concept of belief itself, the concept of subjective probability, and, presumably the most important, the concept of a reason in the sense of a theoretical reason for believing something. Covertly epistemological notions are much more difficult to understand; maybe, they are not epistemological at all. However, a very promising strategy for understanding them is to try to conceive of them as covertly epistemological. One such notion is the concept of objective probability;[1] the concept of explanation is another. A third, very important one is the notion of causation, which has been epistemologically problematic ever since Hume. Finally, there is the notion of truth. Many philosophers believe that there is much to be said for a coherence theory of truth or internal realism; they hold some version of the claim that something for which it is impossible to get a true reason cannot be true, and that truth is therefore covertly epistemological.

Now, if one wants to approach these concepts in a more formal way in order to understand them more clearly and more precisely, the first step will be to try to get a formal grip on epistemology. Here, I am concerned only with this first step.[2] Considering the impressive amount of work in formal epistemology, two general points arise.

The first is very familiar, though it still strikes me as somehow odd; it consists in the fact that formal epistemology, i.e. the formal representation of epistemic states, may be divided into a probabilistic and a deterministic branch (and some things which don't quite fit into the scheme). In a deterministic epistemology, as I call it, one talks about a proposition *being simply believed true or false or neither* by some epistemic subject. The formal machinery established for this works with belief sets, truth in all doxastic alternatives, or similar things well known from epistemic logic.[3] In a probabilistic epistemology, belief is

William L. Harper and Brian Skyrms (eds.), Causation in Decision, Belief Change, and Statistics, II, 105–134.

more finely graded and comes in numerical degrees. The formal machinery appropriate to it is, of course, probability theory.

This dichotomy is naturally prepared for on the intuitive level. All the intuitive notions we have for subjective and objective probability fall on the probabilistic side. Plain belief, of course, belongs to the deterministic side. And so does truth; the simplest reason for this is, I think, that an arbitrary, perhaps uncountable conjunction of truths is still a truth – this being a formal property of truth which cannot be modelled probabilistically. However, the dichotomy is not complete on the intuitive level. The concept of a reason is certainly neutral between the two forms of epistemology. The same holds for the concept of explanation, as we have learned from Hempel, and for the concept of causation, as has been stressed by many who take probabilistic causation seriously.

Of course, one would like to get rid of this dichotomy, i.e. to reduce one side of it to the other; and this can only mean reducing deterministic to probabilistic epistemology. However, this is not so easy, as is highlighted by the famous lottery paradox. Indeed, the different behaviour of conjunction in deterministic and probabilistic formalisms seems to entirely exclude such a reduction. Then, we should do the second best, i.e. we should develop both forms of epistemology as far as possible and then look what we can say about their relations.

Now, however, we have to consider the second point, namely that deterministic epistemology is in a much poorer shape than probabilistic epistemology. One important aspect is that probabilistic epistemology is well entrenched in a behavioral theory, i.e. decision theory; and this is hardly counterbalanced by the fact that a deterministic epistemology can be more easily used in a theory of language.[4] What is more important, however, is that the inner functioning of deterministic epistemology is so much poorer. Usual probabilistic conditionalization and the generalized conditionalization of Jeffrey (1965), ch. 11, give a plausible account of rational epistemic changes. Probability theory also provides a good model for the impact of evidence and counter-evidence on our beliefs, for the weighing of reasons and counter-reasons; it provides, in other words, a good explication for relevance, potential or conditional relevance, and irrelevance in the epistemic sense. As far as I can see, deterministic epistemology can, in the present state, not produce equivalent achievements.

That is precisely what this paper is about; I shall try to raise deterministic epistemology to the level of probabilistic theorizing. More specifically, I shall try to give a more satisfying account of rational changes, i.e. of the dynamics of deterministic epistemic states. It is to be expected, and will become evident, that this brings advance also on the other scores mentioned. Moreover, it will turn out that the problems I am concerned with are in fact present and unsolved at the probabilistic side as well; thus the paper will also add something to probabilistic epistemology.

This being my focus, I greatly simplify my business by proceeding from the obsolete view that belief is a strictly propositional attitude, i.e. that the objects of belief are complete propositions as expressed by eternal sentences. I thereby neglect other serious problems with epistemic states such as the de-re/de-dicto distinction, the fact that belief is most likely neither propositional nor sentential, but something midway, and the observation that belief seems to be as heavily indexical as language itself. But there is no agreed formal epistemology for handling these problems, and our dynamic problem is certainly intricate enough; hence, I comply with that old view and its associated method of possible world talk.

Having thus laid out the general setting, I shall proceed in the following way. First of all, I'd like to keep separate the story I have to tell and the comments relating it to existing ideas and conceptions. My reason for this is not the novelty of the story (only one feature is really new, as far as I know); rather, I wish to do so because: I think that the story is simple and self-contained; I do not want anything read into it which is not explicitly written into it; and the danger of misreading is the greater, the sooner one mixes up this story with similar, but not completely congruent stories. Thus, I defer all comparative remarks to the final Section 8. The story I want to tell starts in Section 2 with a presentation of what I take to be the essentials of the received deterministic conception of epistemic states. In Section 3, I shall state a crucial problem and argue that it cannot be adequately treated within that received conception. In Section 4, I shall introduce my proposal for a solution of this problem i.e. the concept of an ordinal conditional function, and in Sections 5 and 6 the theory of ordinal conditional functions is developed up to a point where it may not be too much to say that this theory offers a genuine qualitative counterpart to probability theory.[5] Finally, Section 7 explains why the whole story also has a considerable bearing for probabilistic epistemology.

2. SIMPLE CONDITIONAL FUNCTIONS

Having made things simple by assuming belief to be propositional, we shall work with the common, technically convenient framework of possible worlds. Thus, throughout this paper, W is to denote a non-empty set of possible worlds (or a sample space, in probabilistic terms).[6] A *proposition* then is just any subset of W.

The most straightforward deterministic representation of an epistemic state is, of course, as a set of propositions, namely those propositions believed true in that state. Will any set of propositions do? No. Usually, it is required, as conditions of rationality, that such a set of propositions be consistent and deductively closed. One might object that this requires an unattainable logical perfection rather than a form of rationality. Indeed; but the logical perfection is already assumed by taking belief to be propositional. For, taking belief to be propositional means that, for any two sentences having the same content, i.e. expressing the same proposition, an epistemic subject should recognize them to have the same content. Thus, it means that epistemic subjects have perfect semantic knowledge which embraces perfect logical knowledge. And given that, the conditions of rationality seem perfectly acceptable; any indication that a subject violates these conditions is also evidence that his semantic knowledge is not perfect.[7]

Formally, these conditions amount to this: If $\mathcal{B}$ is a set of propositions, then $\mathcal{B}$ is consistent iff $\cap\mathcal{B} \neq \emptyset$, and $\mathcal{B}$ is deductively closed iff we have $A \in \mathcal{B}$ whenever there is a $\mathcal{B}' \subseteq \mathcal{B}$ with $\cap\mathcal{B}' \subseteq A$.[8] From this, it follows immediately that, for consistent and deductively closed $\mathcal{B}$, $A \in \mathcal{B}$ iff $\cap\mathcal{B} \subseteq A$. Thus, we can represent an epistemic state simply by a single non-empty proposition C, and the set of propositions believed true in that state is $\{A \mid C \subseteq A\}$. We shall call this proposition C the *net content* of that epistemic state.

If we represent epistemic states simply by their net contents, what can we say about their temporal change? To begin with, it is clear that epistemic changes may have many causes: experiences, forgetfulness, wishful thinking, drugs, etc. And it is also clear that from our armchair position we can at best hope to say something about *rational* epistemic changes on the ground of experience, information and the like. So, suppose that the epistemic state of the subject X at time t has the net content C and that the proposition A represents all the information X gets and accepts between t and t'. What then is the net content C' of

X's epistemic state at t', provided X is not subject to arational influences? We have to distinguish two cases here:

First, consider the case where $C \cap A \neq \emptyset$, i.e. where the new information is compatible with the old beliefs of X. In this case, it is reasonable to assume that $C' \subseteq C \cap A$, since the new information, because of its compatibility with C, does not force X to give up any of his old beliefs. And it is also reasonable to assume that $C \cap A \subseteq C'$; otherwise, X would at t' believe some proposition not implied by his old beliefs and the new information, and there is no good reason for doing so. Thus, rational belief change is in this case characterized by $C' = C \cap A$.

The other case to consider is that $C \cap A = \emptyset$, i.e. that the new information contradicts the old beliefs. This is a very common case; we often learn that we were wrong. And usually, it is an undramatic case; the rearrangement of beliefs usually takes place without much difficulty. However, all attempts to spell out objective principles for the rearrangement of beliefs in this case have failed. The only thing that can at present be confidently said about this case is that X arrives at *some* new epistemic state which includes the belief in A (since A was supposed to be accepted information), i.e. that $\emptyset \neq C' \subseteq A$.

We are thus left with an incomplete account of rational belief change. How can we improve upon the situation? Well, I shall not try to say anything more substantial about the last critical case – as so many have tried to do by invoking such things as lawlike sentences, modal categories, similarity, epistemic importance, informational value, etc., which may appear to be antecedently understandable. Rather, the only thing I shall try to do is to turn what appears to be a partially undetermined process on the surface level of the net contents of epistemic states into a completely determined process on some suitable deeper level. Thus, all the notions introduced in the course of my story are only meant to provide a theoretical substructure to this surface level which derives its meaning exclusively from what it says about the surface level (which I indeed assume to be antecedently understandable). In a sense, we shall only go beneath and not beyond what we have already said. I stress this point, because it seems to involve changing the usual tactics towards our question.

So, what can be done along these lines? Since the above observations about epistemic changes hold for any possible information, we can, as a first reasonable step, define a function which collects all the possible changes of the net contents of epistemic states brought about by all possible informations. Such functions are defined in

DEFINITION 1. The function g is a *simple conditional function (SCF)* iff g is a function from the set of all non-empty subsets of W into the set of all subsets of W such that the following conditions hold for all non-empty A, $B \subseteq W$:

(a) $\emptyset \neq g(A) \subseteq A$,

(b) if $g(A) \cap B \neq \emptyset$, then $g(A \cap B) = g(A) \cap B$.

The interpretation of SCFs is clear: If we use an SCF g for describing X at t, it says that, if A is the information X accepts by $t' > t$, $g(A)$ is the net content of X's epistemic state at t; or briefly: X believes at t B conditional on A iff $g(A) \subseteq B$. This includes that the net content of X's epistemic state at t itself is given by $g(W)$, since the tautological information W leaves X's epistemic state unchanged; hence, X believes B at t iff $g(W) \subseteq B$. An SCF thus provides a *response scheme* to all possible informations.

It is also clear that an SCF should have the properties fixed in Def. 1: The exclusion of the empty set from the domain of an SCF reflects the fact that a contradiction is not an acceptable information. Clause (a) says that, whatever information is accepted, the beliefs remain consistent and include the information. And clause (b) is a natural generalization of what we have said about the case where the new information is compatible with the old beliefs: Our above consideration concluded that, in the present terms, $g(B) = g(W) \cap B$, if $g(W) \cap B \neq \emptyset$; and if we take not, as we did, $g(W)$, but rather the state informed by A, i.e. $g(A)$, as the starting point of that consideration, we just get clause (b).[9]

An SCF is, we understand, a response scheme to all possible informations. Now, a natural further step, which has not been made so far, is to assume that the response scheme which holds for a subject X at some time t is already embodied in the epistemic state of X at t. This means, however, that we give up representing epistemic states simply by their net contents. Rather, we now conceive them as more complicated things representable by SCFs. This is an advance; we can now state a rule for the dynamics of belief which is completely determinate: If the SCF g represents the epistemic state of X at t and if A is the information X accepts between t and t', then X believes B at t' iff $g(A) \subseteq B$ (provided X is not subject to arational influences).

Is this the end of the story? No, for a very simple reason which will be introduced in the next section. Before that, let me introduce an intuitively and technically very useful concept which is equivalent to that of an SCF. Here as well as in all later sections, $\alpha, \beta, \gamma, \ldots, \zeta, \ldots$ will always be used to denote ordinal numbers.

DEFINITION 2. The sequence $(E_\alpha)_{\alpha<\zeta}$ is a *well-ordered partition*, a *WOP* (of W) iff we have for all $\alpha, \beta < \zeta: E_\alpha \neq \emptyset$, $E_\alpha \cap E_\beta = \emptyset$ for $\alpha \neq \beta$, and $\bigcup_{\alpha<\zeta} E_\alpha = W$.

DEFINITION 3. If $(E_\alpha)_{\alpha<\zeta}$ is a WOP and g an SCF, we say that $(E_\alpha)_{\alpha<\zeta}$ *represents* g iff for each non-empty $A \subseteq W$ $g(A) = E_\beta \cap A$, where $\beta = \min\{\alpha | E_\alpha \cap A \neq \emptyset\}$.

THEOREM 1. *Each SCF is represented by exactly one WOP, and each WOP represents exactly one SCF.*

Proof. Let g be an SCF. Define by transfinite recursion: $E_\beta = g(W \backslash \bigcup_{\alpha<\beta} E_\alpha)$. Let ζ be the smallest α for which $E_\alpha = \emptyset$. It is obvious that $(E_\alpha)_{\alpha<\zeta}$ is a WOP. Does it represent g? Yes, as may be seen thus: Let A be a non-empty subset of W and $\beta = \min\{\alpha | E_\alpha \cap A \neq \emptyset\}$. Then we have with the help of clause (b) of Def. 1: $g(A) = g(W \backslash \bigcup_{\alpha<\beta} E_\alpha) \cap A = g(W \backslash \bigcup_{\alpha<\beta} E_\alpha) \cap A = E_\beta \cap A$.

Conversely, let $(E_\alpha)_{\alpha<\zeta}$ be a WOP. Let the function g be defined for all non-empty $A \subseteq W$ as in Def. 3. It is obvious that g then satisfies clause (a) of Def. 1. Now suppose that $g(A) \cap B \neq \emptyset$. This means that $E_\beta \cap A \cap B \neq \emptyset$, where $\beta = \min\{\alpha | E_\alpha \cap A \neq 0\}$. Hence, we also have $\beta = \min\{\alpha | E_\alpha \cap A \cap B \neq \emptyset\}$. This implies that $g(A \cap B) = g(A) \cap B$. Thus, g also satisfies clause (b) of Def. 1, i.e. is an SCF.

Finally, the uniqueness claims of Theorem 1 again are rather obvious. Q.E.D.

A WOP $(E_\alpha)_{\alpha<\zeta}$ is easily interpretable as an *ordering of disbelief* in possible worlds; E_0 contains the possible worlds not disbelieved at all, E_1 contains the least disbelieved worlds, E_2 the second least disbelieved, and so on.[10] The rule for changing beliefs then takes a very simple form: If you now have the ordering $(E_\alpha)_{\alpha<\zeta}$ of disbelief, then you now believe that the true world is among the not disbelieved worlds, i.e. in E_0; thus, E_0 is the net content of your present state. And if you get information A, then you believe that the true world is among the least disbelieved within that information, i.e. in your new net content $E_\beta \cap A$, where $\beta = \min\{\alpha | E_\alpha \cap A \neq \emptyset\}$. What Theorem 1 shows is that response schemes (SCFs) are equivalent to such orderings of disbelief; so we may, and shall indeed, carry through the following considerations in terms of WOPs.

3. A PROBLEM WITH SIMPLE CONDITIONAL FUNCTIONS

So far, we have arrived at conceiving epistemic states as SCFs or WOPs. But there is a problem; the rule for epistemic change we have stated is simply insufficient. In this rule, the old epistemic state was represented by an SCF, but the ensuing epistemic state was still represented in the former way by its net content. This will not do, of course. Having decided to represent epistemic states by SCFs, we must represent *all* epistemic states we are talking of in this way; that is, we must also represent the ensuing state by some SCF, and we must say which SCF that is. The problem becomes pressing, if we consider several successive epistemic changes. The above rule explains the first of these changes; but after that we are back on the surface level of net contents, where we cannot apply the above rule to account for the further changes.

The problem is obvious and grave; but it has received surprisingly little attention. In fact, the only place I found where the problem is explicitly recognized in this way is in Harper (1976, pp. 95ff.), where he tries to solve its probabilistic counterpart with respect to Popper measures.[11] What can we do about it? Well, let's at least try to solve it within our representation of epistemic states. If this should fail, as it will, we shall at least see more clearly what is missing.

It will be intuitively more transparent in this attempt to work with orderings of disbelief, i.e. WOPs. Thus, let the old epistemic state be represented by the WOP

$$E_0, E_1, E_2, \ldots, E_\zeta$$

(which we suppose only for illustrative reasons to have a last term), and let A be the information to be accepted and $\beta = \min\{\alpha \mid E_\alpha \cap A \neq \emptyset\}$. Some new epistemic state ensues which should also be represented by a WOP. Can we determine this new WOP in a reasonable way?

A first proposal might be this: It seems plausible to assume that, after information A is accepted, all the possible worlds in A are less disbelieved than the worlds in $\bar{A}$ (where $\bar{A}$ is the relative complement $W \setminus A$ of A). Further, it seems reasonable to assume that, by getting information only about A, the ordering of disbelief of the worlds within A remains unchanged, and likewise for the worlds in $\bar{A}$. Both assumptions already determine uniquely the new ordering of disbelief; it is given by the sequence

$$E_\beta \cap A, \ldots, E_\zeta \cap A, \quad E_0, \ldots, E_{\beta-1},$$
$$E_\beta \cap \bar{A}, \ldots, E_\zeta \cap \bar{A},$$

where – this is important – all empty terms must still be deleted; otherwise, we wouldn't have a WOP. Wasn't that a quick solution? Well, it isn't a good one. Let me point out three shortcomings:

First, according to this proposal, epistemic changes are not reversible; there is no operation of the specified kind which reinstalls the old ordering of disbelief. In fact, there is in general no way at all, even if we know β, to infer from the new WOP what the old one was. The technical reason for this is just the deletion of empty terms, since after they have been deleted, we no longer know where they have been deleted. However, it is certainly desirable to be able to account for the reversibility of epistemic changes.

Secondly, according to this proposal, epistemic changes are not commutative. If A and B are two logically independent propositions, it is easily checked that getting informed first about A and then about B leads to one WOP, getting informed first about B and then about A leads to another WOP, and getting informed at once about $A \cap B$ leads to still another WOP. This is definitely an inadequacy. To be sure, one wouldn't always want epistemic changes to commute. The two pieces of information may somehow conflict, in which case the order in which they are received may matter. But the normal case is certainly that information just accumulates, and in this case the order of information should be irrelevant. However, according to our proposal it is irrelevant only in trivial cases.

Thirdly, the assumption that, after getting informed about A, all worlds in $\bar{A}$ are more disbelieved than all worlds in A seems too strong. Certainly, the first member, i.e. the net content of the new WOP, must be a subset of A; thus, at least some worlds in A must get less disbelieved than the worlds in $\bar{A}$. But it is utterly questionable whether even the most disbelieved world in A should get less disbelieved than even the least disbelieved world in $\bar{A}$; this could be effected at best by the most certain information.

This last consideration suggests a second proposal. Perhaps one should put only the least disbelieved and not all worlds in A at the top of the new WOP which then looks thus:

$$E_\beta \cap A, \quad E_0, \ldots, E_{\beta-1}, \quad E_\beta \backslash A, \quad E_{\beta+1}, \ldots, E_\zeta.$$

Here again, empty terms still have to be deleted ($E_\beta \backslash A$ may be empty). However, that's no good, either. This proposal does not fare better with respect to the reversibility and commutativity of epistemic changes, as may be easily verified. Moreover, we have now gone to the other extreme. The information A is now treated as only minimally reliable; it is given up as soon as only a single consequence of the things believed together with A, i.e. of $E_\beta \cap A$, turns out to be false.

One may try further. But I think that the case already looks hopeless. There is no good solution to our problem within the confines of SCFs or WOPs. Nevertheless, there are two important conclusions to be drawn from these efforts.

One conclusion is this: In the first proposal the information A was accepted maximally firmly; in the second it was accepted minimally firmly. We considered both extremes undesirable. But then no degree of firmness is the right one for all cases. Rather, the natural consequence is that, in order to specify the new epistemic state, we must say not only which information it is that changes the old state; we must also specify with which firmness this information is incorporated into the new state. This consequence is most important; it means that we have so far neglected a parameter which plays a crucial role in epistemic changes. No wonder that we tried in vain.

The other conclusion is this: We discovered that the reversing of epistemic changes was impossible because of the deletion of empty terms. This suggests that we should generalize the concept of a WOP to the effect that such a partition may contain empty terms. This is what we shall do. Technically, this is a small trick which will, however, make all the difference. Note that this has another important consequence. There may then be two such generalized partitions which order the possible worlds in exactly the same way and which thus differ only by having empty terms at different places. These two partitions should be viewed as two different epistemic states; and this implies that not only the ordering of worlds, but also their relative distances in these partitions are relevant. Mathematically, this means that we have to consider not only the order, but also the arithmetical properties of ordinals.

Now we are well prepared. We only have to adhere to these conclusions. The first conclusion will be developed in Section 5, the second right now.

4. ORDINAL CONDITIONAL FUNCTIONS

It is more convenient to formalize such generalized partitions as functions from possible worlds to ordinals. Moreover, we shall explicitly relativize these functions to a given field of propositions. So far, there was no need for this relativization; but now, when things get more technical, it will prove very useful. The same is done in probability theory, where it is important to compare or relate probability measures on different σ-fields. So, let us define:

DEFINITION 4. Let $\mathcal{A}$ be a complete field of propositions over W (i.e. a non-empty set of subsets of W closed under complementation and arbitrary union and intersection). Then we call κ an *$\mathcal{A}$-measurable ordinal conditional function* (*$\mathcal{A}$-OCF*), if and only if κ is a function from W into the class of ordinals such that $\kappa^{-1}(0) \neq \emptyset$ and for all atoms[12] A of $\mathcal{A}$ and all $w,w' \in A$ $\kappa(w) = \kappa(w')$. Moreover, we define for any $A \in \mathcal{A}\backslash\{\emptyset\}$ $\kappa(A) = \min\{\kappa(w) \mid w \in A\}$.[13]

It is obvious that OCFs generalize WOPs and thus SCFs. The measurability condition is also obvious; it demands that an $\mathcal{A}$-OCF does not discriminate possible worlds which are not discriminated in $\mathcal{A}$.

Two simple observations will be permanently used:

THEOREM 2. *Let κ be an $\mathcal{A}$-OCF. Then we have*

(a) *for each $A \in \mathcal{A}\backslash\{\emptyset,W\}$, $\kappa(A) = 0$ or $\kappa(\bar{A}) = 0$ or both,*

(b) *for all $A,B \in \mathcal{A}\backslash\{\emptyset\}$, $\kappa(A \cup B) = \min\{\kappa(A), \kappa(B)\}$.*

Intuitively, an OCF is not only an ordering, but a *grading of disbelief* in possible worlds. It is clear how such a grading of disbelief is to be understood as a deterministic epistemic state: In state κ, the true world is always believed to be in $\kappa^{-1}(0)$; thus $\kappa^{-1}(0)$ is the net content of the epistemic state κ, and hence the stipulation that $\kappa^{-1}(0) \neq \emptyset$. A is then *believed* in the state κ iff $\kappa^{-1}(0) \subseteq A$, i.e. $\kappa(\bar{A}) > 0$. (Beware: $\kappa(A) = 0$ only means that A is not believed to be false in state κ; and this leaves open the possibility that also $\kappa(\bar{A}) = 0$, i.e. that A is also not believed true in state κ.)

Relative to an OCF κ, we may also introduce degrees of firmness of belief (and thereby slightly reduce the contorted talk of disbelief). If we, for a moment, also allow for negative ordinals, we may say that A *is believed with firmness* α *relative to* κ iff either $\kappa(A) = 0$ and $\alpha = \kappa(\bar{A})$ or $\kappa(A) > 0$ and $\alpha = -\kappa(A)$. Thus, in state κ one believes or disbelieves A iff, respectively, one believes A with positive or negative firmness; firmness 0 means that one is neutral to A. And we might also say that A *is more plausible than* B iff A is believed with greater firmness than B, i.e. iff $\kappa(\bar{A}) > \kappa(\bar{B})$ or $\kappa(A) < \kappa(B)$.[14]

It is clear that the role of taking the minimum corresponds to the role addition has in probability theory; compare the definition of $\kappa(A)$ with the probabilistic formula $P(A) = \Sigma_{w \in A} P(\{w\})$. The two sides of the correspondence differ, however, in a very characteristic way. To put it somewhat metaphorically:

In probability theory, epistemically interpreted, possible worlds have a probability *mass*. They compete for their share of the total mass available; and in epistemic changes these shares get redistributed. Thus, this competition may be conceived as a sort of territorial fight where the parties aim at getting as large as possible. A proposition may then be conceived as a team consisting of its members; and each such team is as weighty and fares as well in this competition as the sum of the masses of its members.

In the theory of OCFs, possible worlds have, by contrast, *grades* of disbelief. They compete for grades, 0 being the top grade above an unending sequence of lower grades; and in epistemic changes their grades will get rearranged. Thus, this competition may be conceived as a sort of race where the parties aim at reaching the top. A proposition may again be conceived as a team consisting of its members; but in this race, each such team is just as good as its best members.[15]

Exactly how do the grades get rearranged in epistemic changes? This is the subject of the next section.

5. CONDITIONALIZATION AND GENERALIZED CONDITIONALIZATION

In what follows I shall make use of the somewhat uncommon *left-sided subtraction* of ordinals[16] which is defined in the following way: Let α

and β be two ordinals with $\alpha \leqslant \beta$; then $-\alpha + \beta$ is to be that uniquely determined ordinal ξ, for which $\alpha + \xi = \beta$.[17]

Moreover, we shall throughout use the following auxiliary concept:

DEFINITION 5. Let κ be an $\mathcal{A}$-OCF and $A \in \mathcal{A}\setminus\{\emptyset\}$. Then the *A-part of* κ is to be that function $\kappa(\cdot|A)$[18] defined on A for which for all $w \in A$ $\kappa(w|A) = -\kappa(A) + \kappa(w)$. For $B \in \mathcal{A}$ with $A \cap B \neq \emptyset$ we also define $\kappa(B|A) = \min\{\kappa(w|A)|w \in A \cap B\} = -\kappa(A) + \kappa(A \cap B)$.

Thus, if $\mathcal{A}' = \{A \cap B|B \in \mathcal{A}\}$, $\mathcal{A}'$ is a complete field of subsets of A, and $\kappa(\cdot|A)$ is an $\mathcal{A}'$-OCF. One might say that the A-part of κ is the restriction of κ to A shifted to 0, i.e. in such a way that the minimum taken is 0. It will soon become clear why I have here chosen the same notation as is used in probability theory.

With the aid of this concept we can define the notion central to the dynamics of epistemic states:

DEFINITION 6. Let κ be an $\mathcal{A}$-OCF, $A \in \mathcal{A}\setminus\{\emptyset,W\}$, and α an ordinal. Then $\kappa_{A,\alpha}$ is to be that $\mathcal{A}$-OCF for which

$$\kappa_{A,\alpha}(w) = \begin{cases} \kappa(w|A), & \text{if } w \in A \\ \alpha + \kappa(w|\bar{A}), & \text{if } w \in \bar{A}\end{cases}.$$

We call $\kappa_{A,\alpha}$ the *A,α-conditionalization of* κ.

Thus, the A,α-conditionalization of κ is the union of the A-part of κ and of the $\bar{A}$-part of κ shifted up by α grades. Trivially, we have $\kappa_{A,\alpha}(A) = 0$ and $\kappa_{A,\alpha}(\bar{A}) = \alpha$; hence, A is believed in $\kappa_{A,\alpha}$ with firmness α. By having introduced the parameter α, we have now taken account of the first conclusion of Section 3.

Def. 6 conforms to the intuitive requirement that getting informed only about A does not change the epistemic state restricted to A, or $\bar{A}$, i.e. the *grading* of disbelief within A, or $\bar{A}$. In other words, the A,α-conditionalization of κ leaves the A-part as well as the $\bar{A}$-part of κ unchanged; they are only shifted in relation to one another. Thereby, we have finally also made use of and given meaning to the relative distances of possible worlds in an OCF, as was implied by our second conclusion of Section 3.

The failure of WOPs may now be seen to have a simple mathematical reason: it's just that the set of all WOPs is not closed under all the above shiftings; therefore, no reasonable conditionalization could be defined for them. With the OCFs, this problem disappears; the class of all OCFs *is* closed under all these shiftings.[19]

The A,α-conditionalization of κ should not always be interpreted as the change of κ which results from obtaining the information A with positive firmness. There are two exceptional cases. For the first case, suppose that $\kappa(\bar{A}) = \beta > 0$; thus, A is believed already in κ. Now, if $\alpha = \beta$, there is no change at all; if $\alpha > \beta$, then one has got additional reason for A whereby the belief in A is strengthened; and if $\alpha < \beta$, then one has got some reason against A whereby the belief in A is weakened, though not destroyed. The second case is the A,0-conditionalization of κ. This may best be described as the neutralization of A and $\bar{A}$, since in $\kappa_{A,0}$ neither A nor $\bar{A}$ is believed. In both cases, it would be inappropriate to say that one was informed about A. But the epistemic changes described in them may certainly be found in reality and are thus properly covered by Def. 6.[20,21]

The problems we had with our proposals in Section 3 no longer trouble us. Of course, epistemic changes according to Def. 6 are reversible:

THEOREM 3. *Let κ be an $\mathcal{A}$-OCF and $A \in \mathcal{A}\setminus\{\emptyset, W\}$ such that $\kappa(A) = 0$ and $\kappa(\bar{A}) = \beta$. Then we have $(\kappa_{A,\alpha})_{A,\beta} = (\kappa_{\bar{A},\alpha})_{A,\beta} = \kappa$.*

Moreover, accumulating information commutes. Here, as in the sequel, we shall say that two ordinals α and β *commute* iff $\alpha + \beta = \beta + \alpha$.

THEOREM 4. *Let κ be an $\mathcal{A}$-OCF and $A,B \in \mathcal{A}\setminus\{\emptyset, W\}$ such that $\kappa(A \cap B) = \kappa(A \cap \bar{B}) = \kappa(\bar{A} \cap B) = 0$, and let α and β be two commuting ordinals. Then we have $(\kappa_{A,\alpha})_{B,\beta} = (\kappa_{B,\beta})_{A,\alpha}$.*

Proof. Set $C_1 = A \cap B$, $C_2 = A \cap \bar{B}$, $C_3 = \bar{A} \cap B$, and $C_4 = \bar{A} \cap \bar{B}$, and for $n = 1,\ldots,4$ $\kappa(C_n) = a_n$, $\kappa_{A,\alpha}(C_n) = b_n$, $(\kappa_{A,\alpha})_{B,\beta}(C_n) = c_n$, $\kappa_{B,\beta}(C_n) = d_n$, and $(\kappa_{B,\beta})_{A,\alpha}(C_n) = e_n$. It suffices to show that $c_n = e_n$ for $n = 1,\ldots,4$: We have assumed that $a_1 = a_2 = a_3 = 0$. By Def. 6 we now get:

$$b_1 = 0, \quad b_2 = 0, \quad b_3 = \alpha, \quad b_4 = \alpha + a_4, \quad \text{and}$$
$$d_1 = 0, \quad d_2 = \beta, \quad d_3 = 0, \quad d_4 = \beta + a_4.$$

Again applying Def. 6, we get from this:

$$c_1 = 0, \quad c_2 = \beta, \quad c_3 = \alpha, \quad c_4 = \beta + \alpha + a_4, \quad \text{and}$$
$$e_1 = 0, \quad e_2 = \beta, \quad e_3 = \alpha, \quad e_4 = \alpha + \beta + a_4.$$

Thus $c_n = e_n$ for $n = 1,2,3$, and also $c_4 = e_4$, since α and β commute.

Q.E.D.

The conclusion of Theorem 4 holds also under more general conditions. These, however, are not so illuminating as to justify the clumsy calculations needed.

We may further generalize our topic. As is well known, Jeffrey (1965, ch. 11) made a substantial contribution to the dynamics of probabilistic epistemic states by discovering generalized conditionalization. There, a probability measure P is conditionalized not by some proposition A, but rather by a probability measure Q on some set of propositions. Q represents here some new state of information with respect to these propositions, and the generalized conditionalization of P by Q describes how the total epistemic state P changes because of this new state of information. Nobody seems to have even thought of doing the same for deterministically conceived epistemic states; but here, the parallel extends in quite a natural way:

DEFINITION 7. Let $\mathcal{B}$ be a complete subfield of $\mathcal{A}$, κ an $\mathcal{A}$-OCF, and λ a $\mathcal{B}$-OCF. Then κ_λ is to be that $\mathcal{A}$-OCF for which for all atoms B of $\mathcal{B}$ and all $w \in B$ $\kappa_\lambda(w) = \lambda(B) + \kappa(w|B)$. We call κ_λ the *λ-conditionalization of* κ.

Def. 6 is only a special case of Def. 7:

THEOREM 5. *Let κ be an $\mathcal{A}$-OCF, $A \in \mathcal{A}\backslash\{\emptyset,W\}$, and λ that $\{\emptyset,A,\bar{A},W\}$-measurable* OCF *for which*

$$\lambda(w) = \begin{cases} 0 & \textit{for} \quad w \in A, \\ \alpha & \textit{for } w \in A. \end{cases}$$

Then, $\kappa_\lambda = \kappa_{A,\alpha}$.

Of course, generalized conditionalization is reversible, too:

THEOREM 6. *Let $\mathcal{B}$, κ, and λ be as in Def.* 7; *and let κ' be the $\mathcal{B}$-measurable coarsening of κ defined by $\kappa'(w) = \kappa(B)$ for all atoms B of $\mathcal{B}$ and all $w \in B$. Then $(\kappa_\lambda)_{\lambda'} = \kappa$.*

With the aid of Def. 7 we can state our most general rule for rational epistemic change: Let X's epistemic state at time t with respect to the field $\mathcal{A}$ of propositions be represented by the $\mathcal{A}$-OCF κ. Suppose further that the experiences between t and t' directly affect only X's attitude towards propositions in the field $\mathcal{B}$ and cause him to adopt the $\mathcal{B}$-OCF λ as epistemic state with respect to $\mathcal{B}$. Then κ_λ represents X's epistemic state at t' with respect to $\mathcal{A}$ (provided X is not subject to arational influences).

This formulation of the rule brings out a fact which seems by now to be well accepted in epistemology in general. It was realized in probabilistic epistemic modelling with Jeffrey's generalized conditionalization (this was its revolutionary point), but it does not seem to have been clearly recognized in deterministic epistemic modelling: I mean the fact that what is described by rules of epistemic change are never rational inner reactions to outward circumstances or happenings, but always rational adjustments of the overall epistemic state to inner epistemic changes in particular quarters; how these initial epistemic changes come about is in any case a matter to which a rationality assessment cannot be reasonably applied and which therefore falls outside the scope of investigations like this one. This fact is formally mirrored, here as in Jeffrey, by the fact that epistemic states, probability measures or OCFs, are conditionalized by things of their own kind; talking of conditionalization by propositions (or events), albeit technically correct, has been intuitively very misleading.

6. INDEPENDENCE AND CONDITIONAL INDEPENDENCE

Related to conditionalization, there is another important topic in probability theory in particular, but also in epistemology in general: namely dependence and independence. I know of no reasonable definition of independence for deterministic representations of epistemic states. Logical independence will not do, of course, since almost everything is logically independent of almost everything. The best we can do within the domain of SCFs is to say that A is epistemically independent of B relative to the SCF g, if and only if $g(B) \subseteq A$ iff $g(\bar{B}) \subseteq A$ and $g(B) \subseteq \bar{A}$ iff $g(\bar{B}) \subseteq \bar{A}$, i.e. iff acceptance of B, or of $\bar{B}$, does not matter to whether A or $\bar{A}$ or neither is believed. However, this implies, for example, that each A believed true in state g is independent of each B

believed neither true nor false in g; and this is certainly much too much independence.

Not surprisingly, there is no problem with independence with respect to OCFs:

DEFINTION 8. Let κ be an $\mathcal{A}$-OCF and $\mathcal{B}$ and $\mathcal{C}$ two complete subfields of $\mathcal{A}$. Then $\mathcal{C}$ is *independent of* $\mathcal{B}$ *with respect to* κ iff for all atoms B of $\mathcal{B}$ and all atoms C of $\mathcal{C}$ $B \cap C \neq \emptyset$ and $\kappa(B \cap C) = \kappa(B) + \kappa(C)$. $\mathcal{B}$ *and* $\mathcal{C}$ are *independent* (with respect to κ) iff $\mathcal{C}$ is independent of $\mathcal{B}$ and $\mathcal{B}$ is independent of $\mathcal{C}$. Moreover, if $A,B \in \mathcal{A}\setminus\{\emptyset,W\}$, A is *independent of* B (with respect to κ) iff $\{\emptyset,A,\bar{A},W\}$ is independent of $\{\emptyset,B,\bar{B},W\}$, and A *and* B are *independent* (with respect to κ) iff A is independent of B and B is independent of A.

Def. 8 copies probabilistic independence concepts as far as possible. Note that independence with respect to OCFs need not be symmetric, simply because addition of ordinals is not commutative; therefore the distinction between "A is independent of B" and "A and B are independent".

Independence so defined has the properties we would expect.

THEOREM 7. *If* $\mathcal{C}$ *is independent of* $\mathcal{B}$ *with respect to* κ, *then for all* $B \in \mathcal{B}\setminus\{\emptyset\}$ *and all* $C \in \mathcal{C}\setminus\{\emptyset\}$:

$$\kappa(B \cap C) = \kappa(B) + \kappa(C).$$

Proof. Let $B' \in \mathcal{B}\setminus\{\emptyset\}$ *and* $C' \in \mathcal{C}\setminus\{\emptyset\}$. Let further $\mathcal{B}'$ be the set of atoms of $\mathcal{B}$ which are subsets of B' and $\mathcal{C}'$ the set of atoms of $\mathcal{C}$ which are subsets of C'. Hence, $B'=\cup\mathcal{B}'$ and $C'=\cup\mathcal{C}'$, and moreover, $\kappa(B') = \min\{\kappa(B)|B \in \mathcal{B}'\}$ and $\kappa(C') = \min\{\kappa(C)|C \in \mathcal{C}'\}$. Then we have

$$\begin{aligned}\kappa(B' \cap C') &= \min\{\kappa(B \cap C)|B \in \mathcal{B}', C \in \mathcal{C}'\} = \\ &= \min\{\kappa(B) + \kappa(C)|B \in \mathcal{B}', C \in \mathcal{C}'\} = \\ &= \min\{\kappa(B)|B \in \mathcal{B}'\} + \min\{\kappa(C)|C \in \mathcal{C}'\} = \\ &= \kappa(B') + \kappa(C').\end{aligned}$$

Q.E.D.

The converse of Theorem 7 is obviously true. An immediate consequence of Theorem 7 and Def. 5 and 7 is

THEOREM 8. *The following three assertions are equivalent:*

(a) $\mathcal{C}$ *is independent of* $\mathcal{B}$ *with respect to* κ,

(b) *for all* $B \in \mathcal{B}\backslash\{\emptyset\}$ *and* $C \in \mathcal{C}\backslash\{\emptyset\}$ $\kappa(C|B) = \kappa(C)$ *holds true,*

(c) *for each* $\mathcal{B}$-OCF λ *and each* $C \in \mathcal{C}\backslash\{\emptyset\}$ $\kappa_\lambda(C) = \kappa(C)$ *holds true.*

Theorem 8 particularly clearly shows the intuitive adequacy of Def. 8.

The parallel to probability theory may be extended further. In probability theory, one also defines independence for families of subfields. This can be done here as well.

DEFINITION 9. Let $(\mathcal{B}_\alpha)_{\alpha<\beta}$ be a sequence of complete subfields of $\mathcal{A}$ and κ an $\mathcal{A}$-OCF. Then $(\mathcal{B}_\alpha)_{\alpha<\beta}$ is called *independent with respect to* κ iff for all atoms B_α of $\mathcal{B}_\alpha$ $(\alpha < \beta)$ $\bigcap_{\alpha<\beta} B_\alpha \neq \emptyset$ and $\kappa(\bigcap_{\alpha<\beta} B_\alpha) = \Sigma_{\alpha<\beta}\,\kappa(B_\alpha)$.

The connection to Def. 8 is stated in

THEOREM 9. $(\mathcal{B}_\alpha)_{\alpha<\beta}$ *is independent iff for all* $\gamma < \beta$ *the complete field generated by* $\bigcup_{\gamma\leqslant\alpha<\beta} \mathcal{B}_\alpha$ *is independent of the complete field generated by* $\bigcup_{\alpha<\gamma} \mathcal{B}_\alpha$.

Proof. Define $\mathcal{C}_\gamma$ and $\mathcal{D}_\gamma$ to be, respectively, the complete field generated by $\bigcup_{\alpha<\gamma} \mathcal{B}_\alpha$ and $\bigcup_{\gamma\leqslant\alpha<\beta} \mathcal{B}_\alpha$. Now suppose first that for all atoms B_α of $\mathcal{B}_\alpha$ $(\alpha < \beta)$ $\kappa(\bigcap_{\alpha<\beta} B_\alpha) = \Sigma_{\alpha<\beta}\,\kappa(B_\alpha)$. This implies that for all atoms B_α of $\mathcal{B}_\alpha$ $(\alpha < \gamma)$ $\kappa(\bigcap_{\alpha<\gamma} B_\alpha) = \Sigma_{\alpha<\gamma}\,\kappa(B_\alpha)$, and similarly, that for all atoms B_α of $\mathcal{B}_\alpha$ $(\gamma \leqslant \alpha < \beta)$ $\kappa(\bigcap_{\gamma\leqslant\alpha<\beta} B_\alpha) = \Sigma_{\gamma\leqslant\alpha<\beta}\,\kappa(B_\alpha)$. Thus, we have $\kappa(\bigcap_{\alpha<\beta} B_\alpha) = \kappa(\bigcap_{\alpha<\gamma} B_\alpha) + \kappa(\bigcap_{\gamma\leqslant\alpha<\beta} B_\alpha)$, and this means that $\mathcal{D}_\gamma$ is independent of $\mathcal{C}_\gamma$.

Conversely, suppose that for all $\gamma < \beta$ $\mathcal{D}_\gamma$ is independent of $\mathcal{C}_\gamma$. For $\gamma = 1$, this says that for all atoms B_α of $\mathcal{B}_\alpha$ $(\alpha < \beta)$ $\kappa(\bigcap_{\alpha<\beta} B_\alpha) = \kappa(B_0) + \kappa(\bigcap_{1\leqslant\alpha<\beta} B_\alpha)$. This implies in particular that for all atoms B_α of $\mathcal{B}_\alpha$ $(\alpha = 0,1)$ $\kappa(B_0 \cap B_1) = \kappa(B_0) + \kappa(B_1)$. For $\gamma = 2$, we therefore get that for all atoms B_α of $\mathcal{B}_\alpha$ $(\alpha < \beta)$ $\kappa(\bigcap_{\alpha<\beta} B_\alpha) = \kappa(B_0 \cap B_1) + \kappa(\bigcap_{2\leqslant\alpha<\beta} B_\alpha) = \kappa(B_0) + \kappa(B_1) + \kappa(\bigcap_{2\leqslant\alpha<\beta} B_\alpha)$. Continuing this line of reasoning by transfinite induction till β then leads to the desired result. Q.E.D.

An immediate consequence of Theorem 9 is

THEOREM 10. *Let* $(\mathcal{B}_\alpha)_{\alpha<\beta}$ *be independent. Let* $(\Gamma_\gamma)_{\gamma<\delta}$ *be a partition of* $\{\alpha|\alpha < \beta\}$ *such that we have for all* $\gamma,\gamma' < \delta$: *if* $\gamma < \gamma'$, *then*

$\alpha < \alpha'$ for all $\alpha \in \Gamma_\gamma$ and $\alpha' \in \Gamma_{\gamma'}$. Let finally $\mathcal{C}_\gamma$ be the complete field generated by $\bigcup_{\alpha \in \Gamma_\gamma} B_\alpha$. Then the sequence $(\mathcal{C}_\gamma)_{\gamma<\delta}$ is also independent.

In probability theory, the corresponding theorem is known as the theorem of the composition of independent fields.

As a last topic, let me take up conditional independence. It is well known that conditional independence is central for a probabilistic theory of causality. Thus, this topic will become important, when one turns to deterministic theories of causality. Here, however, I take it up only for demonstrating the parallel between probability measures and OCFs a bit further.

DEFINITION 10. Let $\mathcal{B}$ and $\mathcal{C}$ be two complete subfields of $\mathcal{A}$, κ an $\mathcal{A}$-OCF, and $A \in \mathcal{A} \setminus \{\emptyset\}$. Then $\mathcal{C}$ is *independent of $\mathcal{B}$ conditional on A* (or *given A*) *with respect to* κ iff for all atoms B of $\mathcal{B}$ and all atoms C of $\mathcal{C}$ with $A \cap B \cap C \neq \emptyset$ $\kappa(B \cap C|A) = \kappa(B|A) + \kappa(C|A)$. If $\mathcal{D}$ is another complete subfield of $\mathcal{A}$, then $\mathcal{C}$ is *independent of $\mathcal{B}$ conditional on $\mathcal{D}$* (or *given $\mathcal{D}$*) (with respect to κ) iff for each atom D of $\mathcal{D}$ $\mathcal{C}$ is independent of $\mathcal{B}$ given D. Further phrases may be defined in analogy to Def. 8.

The intuitive interpretation of Def. 10 should be clear and is supported by the fact that Theorems 7 and 8 hold correspondingly for conditional independence. The following theorems are more interesting; the expression "$\mathcal{B} + \mathcal{C}$" used in them is meant to denote the complete field generated by $\mathcal{B} \cup \mathcal{C}$.

THEOREM 11. *Let $\mathcal{B}$, $\mathcal{C}$, $\mathcal{D}$, and $\mathcal{E}$ be four complete subfields of $\mathcal{A}$. Suppose that $\mathcal{C}$ is independent of $\mathcal{B}$ given $\mathcal{D} + \mathcal{E}$ and that $\mathcal{D}$ is independent of $\mathcal{B}$ given $\mathcal{E}$. Then $\mathcal{C} + \mathcal{D}$ is independent of $\mathcal{B}$ given $\mathcal{E}$.*

Proof. Let B, C, D, and E be variables for atoms of $\mathcal{B}$, $\mathcal{C}$, $\mathcal{D}$, and $\mathcal{E}$, respectively. The first assumption says that for all B, C, D, and E with $B \cap C \cap D \cap E \neq \emptyset$:

$\kappa(B \cap C|D \cap E) = \kappa(B|D \cap E) + \kappa(C|D \cap E)$, i.e. by Def. 5,
$-\kappa(D|E) + \kappa(B \cap C \cap D|E) =$
$= -\kappa(D|E) + \kappa(B \cap D|E) + (-\kappa(D|E) + \kappa(C \cap D|E))$, i.e.
$\kappa(B \cap C \cap D|E) = \kappa(B \cap D|E) + (-\kappa(D|E) + \kappa(C \cap D|E))$.

The second assumption states that for all B, D, and E with $B \cap D \cap E \neq \emptyset$:

$$\kappa(B \cap D|E) = \kappa(B|E) + \kappa(D|E).$$

The last two equations together yield that for all B, C, D, and E with $B \cap C \cap D \cap E \neq \emptyset$

$$\kappa(B \cap C \cap D|E) = \kappa(B|E) + \kappa(C \cap D|E);$$

and that's what we had to prove. Q.E.D.

In the same way, the result symmetric to Theorem 11 may be proved:

THEOREM 12. *If $\mathcal{B}$ is independent of $\mathcal{C}$ given $\mathcal{D} + \mathcal{E}$ and independent of $\mathcal{D}$ given $\mathcal{E}$, then $\mathcal{B}$ is independent of $\mathcal{C} + \mathcal{D}$ given $\mathcal{E}$.*

Moreover, we have

THEOREM 13. *If $\mathcal{B}$ is independent of $\mathcal{C} + \mathcal{D}$ given $\mathcal{E}$ and independent of $\mathcal{C} + \mathcal{E}$ given $\mathcal{D}$, then $\mathcal{B}$ is also independent of $\mathcal{C} + \mathcal{D} + \mathcal{E}$ given $\mathcal{D} \cap \mathcal{E}$.*

Proof. Let B, C, D, and E be as in the proof of Theorem 11, and let F be a variable for the atoms of $\mathcal{D} \cap \mathcal{E}$ (which, to be sure, is also a complete field). The first premise says that for all B, C, D, E, and F with $B \cap C \cap D \cap E \neq \emptyset$ and $D, E \subseteq F$:

$$\begin{aligned}
&\kappa(C \cap D \cap B|E \cap F) = \\
&= \kappa(C \cap D|E \cap F) + \kappa(B|E \cap F), \text{ i.e. by Def. 5,} \\
&-\kappa(E|F) + \kappa(C \cap D \cap E \cap B|F) = \\
&= -\kappa(E|F) + \kappa(C \cap D \cap E|F) + (-\kappa(E|F) + \kappa(E \cap B|F)), \text{ i.e.} \\
(1)\quad &\kappa(C \cap D \cap E \cap B|F) = \\
&= \kappa(C \cap D \cap E|F) + (-\kappa(E|F) + \kappa(E \cap B|F)).
\end{aligned}$$

Likewise, the second premise says that for all B, C, D, E, and F with $B \cap C \cap D \cap E \neq \emptyset$ and $D, E \subseteq F$:

$$\begin{aligned}
&\kappa(C \cap E \cap B|D \cap F) = \\
&= \kappa(C \cap E|D \cap F) + \kappa(B|D \cap F), \text{ i.e. as before}
\end{aligned}$$

$$\begin{aligned}
(2)\quad &\kappa(C \cap D \cap E \cap B|F) = \\
&= \kappa(C \cap D \cap E|F) + (-\kappa(D|F) + \kappa(D \cap B|F)).
\end{aligned}$$

(1) and (2) imply that for all D, $E \subseteq F$

$$-\kappa(E|F) + \kappa(E \cap B|F) = -\kappa(D|F) + \kappa(D \cap B|F).$$

This in turn implies that for all atoms E,E' of $\mathscr{E}$ with $E,E' \subseteq F$

$$(3) \qquad -\kappa(E|F) + \kappa(E \cap B|F) = -\kappa(E'|F) + \kappa(E' \cap B|F).$$

Now, there must be an atom E_0 of $\mathscr{E}$ with $E_0 \subseteq F$ such that $\kappa(E_0|F) = 0$, since $0 = \kappa(F|F) = \min\{\kappa(E|F)|E \subseteq F\}$. Thus, we have, using (3), $\kappa(E_0 \cap B|F) = \min\{\kappa(E \cap B|F)|E \subseteq F\} = \kappa(B|F)$. Using (3) once more, this yields that for all $E \subseteq F$

$$-\kappa(E|F) + \kappa(E \cap B|F) = \kappa(B|F).$$

Substituting this result in (1), we finally get

$$\kappa(C \cap D \cap E \cap B|F) = \kappa(C \cap D \cap E|F) + \kappa(B|F)$$

for all B, C, D, E, and F with $B \cap C \cap D \cap E \neq \emptyset$ and D, $E \subseteq F$.

Q.E.D.

The assertion symmetric to Theorem 13 does not necessarily hold.

These theorems are as analogous to probabilistic theorems as can be.[22] Here I would like to end for the time being. I think there can be no doubt that OCFs are vastly superior to SCFs or WOPs.

7. CONNECTIONS WITH PROBABILITY THEORY

So far, all this has been a story wholly within deterministic epistemology. But there is in fact an exact probabilistic duplicate of our story progressing from net contents to OCFs. The probabilistic counterparts to our net contents are probability measures. With net contents we had the problem that we could say nothing about the new net content resulting from information incompatible with the old net content; this problem induced us to introduce the SCFs. The corresponding problem is that probabilities conditional on propositions having probability 0 are not defined in standard probability theory; we can say nothing about the new probability measure resulting from information having probability 0 in the old epistemic state. One solution to this problem, perhaps the most prominent one, consists in introducing Popper measures.[23] These are indeed the probabilistic counterparts to our SCFs; it is known that it is an SCF which, if adapted to the algebraic framework of probability theory (which operates with σ-fields instead of complete fields), represents the 0-1-structure of a Popper measure P, i.e. the relation $\{\langle A,B\rangle \mid P(B \mid A) = 1\}$.[24] This means, however, that Popper measures are as insufficient for a dynamic theory of epistemic states as SCFs are; this was very clearly pointed out by Harper (1976, pp. 95f). Hence, the probabilistic story calls for continuation, too. It is quite obvious what this should look like; just define probabilistic counterparts to OCFs which would be something like functions from propositions to ordered pairs consisting of an ordinal and a real between 0 and 1. I won't now pursue this in technical detail, since this fusion of probability theory and the theory of OCFs appears to me to be fairly straightforward. But the advantage of such probabilified OCFs over Popper measures is quite clear; it is the same as that of OCFs over SCFs.

One point, however, is still open; we do not yet have an *explanation* of why OCFs behave so much like probability measures (which is, of course, not given by the proposed fusion of probability theory and the theory of OCFs). But this explanation may be made, I think, along the following lines within the framework of nonstandard probability theory:[25] Let P be a nonstandard probability measure for which there is an infinitesimal i such that for each A $P(A)$ is of the same order as i^n for some (nonstandard) natural number n (i.e. $P(A)/i^n$ is finite, but not infinitesimal). Now define $\kappa(B \mid A) = n$ iff $P(B \mid A)$ is of the same order as i^n. Then κ is like an OCF within this framework. Indeed, we have thereby defined a homomorphism from a class of nonstandard probability measures onto the class of (nonstandard) OCFs which maps, first, addition of probabilities into taking the minimum of OCF-values and which maps, secondly, multiplication and division of probabilities into addition and subtraction of OCF-values. More specifically, whenever A and B are independent according to P, they are so according to κ; and for the κ so defined, we have $\kappa(B \mid A) = \kappa(A \cap B) - \kappa(A)$. This would explain why OCFs obey the same laws as probability measures concerning independence and conditionalization.[26]

8. DISCUSSION

I see three points where the foregoing story should be related to the actual state of discussion.

The first point is that all concepts introduced in Section 2 are, of course, absolutely standard. SCFs are better known under the label of (class) selection functions, which play a central role in conditional logic; in this context, a number of slightly different concepts of a selection function have been proposed, and it is well known that nearly every semantics for conditional logic is based on some such concept.[27] Moreover, if $(E_\alpha)_{\alpha<\zeta}$ is a WOP, then the sequence $(\cup_{\alpha\leq\beta} E_\alpha)_{\beta<\zeta}$ is a (universal) system of similarity spheres (at one possible world) in the sense of David Lewis. (In general, a system of spheres need not to be well-ordered, of course.) Thus, Theorem 1 can be already found in Lewis (1973, at pp. 58f.) and in other places.

Why, then, did I define the SCFs in the way I did? Well, I have stated my reasons for doing so fully in Section 2. These reasons are debatable, but I have the impression that the slight differences between the various concepts of a selection function are motivated rather by differing opinions about conditional logic than about the dynamics of belief; and I was exclusively concerned with the latter which must not be mixed up with the former. (That was one reason why I deferred this comment.) To be sure, I completely side with Ernest W. Adams, Brian Ellis, Peter Gärdenfors, and others in maintaining that the various uses of the conditional can only be correctly and uniformly understood by relating them to a dynamic theory of epistemic states. But this relation is, I think, not yet sufficiently understood.

To be a bit more specific: If one accepts something like the straight thesis that the sentence "if *A*, then *B*" is accepted in (or true relative to) some epistemic state if and only if *B* is accepted in the revision of that state by *A*,[28] then one is bound to strain one or other side of this biconditional. A clear case, in my view, is provided by the very common causal conditionals. For, as I in effect argue in my (1983), if the concept of revising epistemic states is only to tell how beliefs change, then, according to this thesis, the conditional "if *A*, then *B*" only states something about the evidential relations between *A* and *B*, i.e. about *A*'s being a reason for *B*, and thus does not yet express a causal relation between *A* and *B*. But let's not go further into this; my remark should only show why I want to confine myself to the dynamics of epistemic states and to leave aside the complicated relations to conditional logic.

The second point is this: If the central problem stated in Section 3 has been known at least since Harper (1976), what has been done to solve it? Surprisingly, not very much; and one reason for this is, it seems to me, that the issue has been obscured by what I have just complained about, i.e. by not clearly separating the dynamics of belief and conditional logic. In fact, I have found only three ideas which are addressed to this issue or can be so understood; and I shall deal with them, for the sake of brevity, only at a strategic level:

The first idea is that our problem of accounting for iterated belief changes appears to be analogous to the problem of providing a semantics for a language with iterated conditionals. The standard solution to the latter problem is to associate an SCF, a selection function, a system of similarity spheres, or whatever with *every* possible world (so that each conditional sentence has again a set of possible worlds as its truth condition). There is no need now to assess the semantic problem and its solution, though I always had the impression that in iterated intensional constructions the syntactic horse bolts with the semantic rider. The main point is that I don't see how the seeming analogy could be brought to bear; for, how should such a function from possible worlds to SCFs or whatever be interpreted as an epistemic state?

A second related idea is this: Enrich the language in which propositions are expressed by a conditional and thus by conditional sentences and propositions, and then exploit this new structural richness of the epistemic objects for a solution of our problem. This is, very roughly, the strategy applied by Harper (1976, pp. 95ff.). Ellis (1979, pp. 53ff.) and Gärdenfors (1979 and 1981, sect. II and III), seem to endorse it as well. However this strategy is brought to work in detail, it seems to be wrong from the start for two reasons: Our problem with SCFs shows, one should think, that it is the characterization of epistemic states as SCFs and not the structure of the epistemic objects which is too poor; one would expect that a dynamic theory of epistemic states does not force us to make special assumptions about the underlying structure of the epistemic objects. So, this strategy seems to focus on the wrong point (whereas our OCFs conform to this expectation). Moreover, there is the problem of how the conditional is interpreted within this strategy. In order to keep within the spirit of their approach, Harper and the others want to interpret it in terms of the dynamics of belief so far elaborated. This, however, amounts in fact to assuming second or higher order epistemic states which are partially about propositions describing properties of lower order epistemic states. Interesting as this may be, this move is uncalled for; one would expect that the problem with SCFs can be solved strictly at the level of first-order epistemic states. Thus, this second idea seems to be an unconvincing mixture of the dynamics of belief and conditional logic.

The third and last approach is found in Gärdenfors (1984). The machinery developed there consists of belief sets, which are essentially equivalent to our net contents, and a relation of epistemic importance between sentences or propositions. With this machinery, Gärdenfors is able to describe successive changes of belief sets and thus gives a solution to our problem with SCFs – *provided* that the relation of epistemic importance is kept fixed. But why should it be so? An ordering of disbelief in our sense does essentially the same job as a belief set plus a relation of epistemic importance (though our respective interpretations of the two things do not match precisely); thus, that relation should be viewed as a part of an epistemic state which may change, too. Gärdenfors' approach therefore seems to me to provide only a restricted solution to our problem.

All this considered, there is enough reason to look for a solution to our problem elsewhere, as I have done in Sections 3–5.

The final point in need of a comment is that our OCFs look rather familiar; our degrees of disbelief seem more or less identical with the degrees of potential surprise in Shackle (1969). Indeed, the similarity is amazing, and the more so as Shackle developed his ideas long ago (before there was any conditional logic) and in quite a different scientific department. Since in particular his intuitive explanation of his functions of potential surprise perfectly fit my OCFs, it may be worthwhile to identify the points of difference, although this comparison is bound to be forced and somewhat unfair just because of the very different setting of his work.

According to Shackle (1969, p. 80),[29] a *function y of potential surprise* (an *FPS*) may be defined to be a function from a given field of propositions into the closed interval [0,1] such that for all propositions A and B

(1) $y(\emptyset) = 1$,
(2) either $y(\mathrm{A}) = 0$ or $y(\bar{A}) = 0$ or both,
(3) $y(A \cup B) = \min\{y(A), y(B)\}$.

1 is the arbitrarily chosen maximal degree of potential surprise which is taken at least by Ø, and (2) and (3) are identical with my Theorem 2 (Sect. 4). Thus, there seem to be hardly any differences between FPSs and OCFs; but there are four.

One point is that OCFs satisfy the generalization of (3) to arbitrary unions; but, as expressed in Notes 8 and 13, I do not attach much importance to this. Since in this generalization min is not weakened to inf, it forces the range of OCFs to be well-ordered; and then ordinals are the natural values for OCFs. Thus, the difference with respect to (3) also accounts for the differing ranges of FPSs and OCFs; but we shall see that there is more to the difference in the ranges.

Another difference is about the maximal degree of potential surprise. I also could have introduced a number larger than any ordinal as the OCF-value for $\emptyset$; but this did not look nice, and so I preferred to make qualifications to the effect that $\emptyset$ is not imported into the domain of an OCF. The important point here is that I therefore do not allow any other proposition to take the maximal value. The reason is that, once a proposition were disbelieved to the maximal degree, it would always be disbelieved to the maximal degree, at least according to my rules of belief change; rational belief change could then no longer be treated within my framework. This was something I wanted to avoid. Shackle, by contrast, makes free use of the maximal degree of potential surprise. And Levi (1980, p. 7) explicitly assigns it to each proposition that is incompatible with what he calls a corpus of knowledge, and he therefore has trouble, e.g. in (1983), with specifying rules for changing such corpora of knowledge.

The essential point is that Shackle has no precise and workable account of conditional degrees of potential surprise, of changes of FPSs, etc. This becomes apparent in his handling of conjunctions. In his (1969, pp. 80ff. and 199ff.), he sticks to the postulate that

$$(4) \qquad y(A \cap B) = \max\{y(A), y(B|A)\}$$

(where I have adapted the notation and where $y(B|A)$ is in fact undefined). In contrast to this, our Definition 5, which is fundamental for our Sections 5 and 6, is equivalent to

$$(5) \qquad \kappa(A \cap B) = \kappa(A) + \kappa(B|A).$$

Shackle has obviously considered accepting something like (5) instead of (4); but he says little about why he finally rejected it. In his (1969, p. 205), he says only that (4) would be simpler and less unrealistic than something like (5).

A final significant difference may be inferred from (1)–(5). Shackle (1969, ch. XV–XVII) clearly intends his FPSs to be measurable on a ratio scale. But it is hard to see precisely how this scale is established and where it is really used; it seems that we may conceive FPSs as purely ordinal concepts.[30] In any case, FPSs as displayed by (1)–(4) are purely ordinal, as may be seen from the exclusive use of mathematical operations like max and min. But if this is so, FPSs correspond to our WOPs (or the functions definable by WOPs according to Note 13). This would mean that the decisive step towards OCFs is perhaps intended, but not really taken by FPSs.

Institut für Philosophie
Universität Regensburg

NOTES

* Many have helped. I am very much indebted: first to Godehard Link for spending a Christmas vacation making various helpful remarks and suggestions; to Wolfgang Stegmüller and Max Drömmer for fruitful discussion; to Kurt Weichselberger for decisive help in Section 7; to Brian Skyrms and Bill Harper for giving me the opportunity to present this paper at their conference; to Peter Gärdenfors for further fruitful discussion; to Isaac Levi for some enlightening controversies and for drawing my attention to the work of Shackle; not last and not least to Joe Lambert for carefully going through the whole stuff with me twice more and almost forgetting dinner about it; to Jeremy Adler for being so kind to descend from investigating Goethe's German to correcting and improving my English; and finally to the Wissenschaftskolleg zu Berlin for giving me the leisure for preparing the final version of this paper.

[1] Which is clearly conceived as covertly epistemological by Lewis (1980), for instance.

[2] In my (1983), I have proposed a way of progressing from the notion of reason to the notion of cause. I have refrained there from introducing the formal machinery developed in this paper; footnote 18 of that paper marks the point where what I there called selection functions and shall call here simple conditional functions should be replaced by the ordinal conditional functions to be defined – for reasons more fully explained here in Section 3.

[3] I am not happy with the term "deterministic epistemology", but I could not find a better one. It derives from the natural and familiar distinction between deterministic and probabilistic causation which, in my opinion, is closely related to the different forms of epistemology.

[4] Think, e.g., of the disquotation principle saying that if *X* sincerely and seriously utters "*p*", then *X believes* that *p*. This is an important, though not generally true linguistic fact; and it is hard to see what a probabilistic version of it could look like.

[5] This is not to be confused with what is ordinarily called qualitative probability which is a relational, comparative concept.

[6] Where I don't at all oppose construing a possible world as a maximal consistent set of sentences of a given langauge, as a valuation of that language, or the like.

[7] This consideration suggests that the idealization of belief as propositional should be overcome not by seeking for a stricter objective individuation of the objects of belief, but by getting a grip on the subjective imperfections of semantic knowledge.

[8] One might argue about whether $\mathcal{B}'$ should here be assumed to be countable or finite or neither. With my definition of deductive closure I have assumed what has been called the generalized consequence principle; cf. e.g. Pollock (1976, pp. 19f.) or Gärdenfors (1981, p. 308). I do so, because I find this principle convincing given the idealization of perfect semantic knowledge and because it makes things technically much simpler. However, as far as I see, everything I say in this paper could be adapted to a weaker assumption without essential complications.

[9] Some, e.g. Lewis (1973, p. 58), prefer to replace the condition that $g(A) \neq \emptyset$ (which is tantamount to universality) by the condition that $A \subseteq B$ and $g(A) \neq \emptyset$ imply $g(B) \neq \emptyset$. However, that's much of a muchness. The only difference is this: With the alternative definition one can prove that there is a $D \subseteq W$ such that $g(A) = \emptyset$ iff $A \subseteq D$. Now alter g by putting $g(A) = A$ for $A \subseteq D$ and leave it otherwise unchanged (thus, there is no change, if $D = \emptyset$); then g complies with our definition. In both cases, accepting an information $A \subseteq D$ leads to complete epistemic collapse. In our case, only the information is then believed; all induction ceases. In the alternative case, everything is then believed (if this makes sense); induction goes crazy. I find our description of that desperate situation a bit more pleasant; besides, SCFs in our sense are more easily generalized to the ordinal conditional functions introduced in Section 4.

[10] Continuously using these negative terms is a somewhat clumsy and contorted mode of expression. But Isaac Levi has convinced me that this is precisely the intuitively appropriate terminology.

[11] I shall say in Section 7 how Popper measures relate to the present subject.

[12] A is an atom of $\mathcal{A}$ iff $A \neq \emptyset$ and there is no $B \in \mathcal{A}$ with $\emptyset \subset B \subset A$. Complete fields of sets are always atomic.

[13] This latter function for propositions is the more important one. The corresponding notion at the level of WOPs is the function assigning to each proposition A the number $\min\{\alpha \mid E_\alpha \cap A \neq \emptyset\}$, which we have frequently used, though not explicitly introduced. Note, by the way, that it is our acceptance of the generalized consequence principle (cf. Note 8) which is in the end responsible for the possibility of reducing the propositional function to a function defined for possible worlds.

[14] All this is a sort of exercise in intuitively interpreting OCFs. Degrees of firmness could also have been introduced relative to WOPs; but this would have been misleading, because, relative to WOPs, numbers have a purely ordinal meaning.

[15] Still, it would be inappropriate to say that only the best members count. Imagine a proposition having only one member with a good grade, the rest being very far behind. Then, if this good member fell back very badly, so would the whole proposition. If, however, the rest were not so bad, the top member could fall back without disastrous consequences for the team. In this sense the rest matter, too.

[16] It would be a natural idea to restrict the range of OCFs to the set of natural numbers. In fact, much of the following could thereby be simplified since usual arithmetic is simpler than the arithmetic of ordinals. For the sake of formal generality I do not impose this restriction. But larger ranges may also be intuitively needed. For example, it is tempting to use OCFs with larger ranges to represent the stubbornness with which some beliefs are held in the face of seemingly arbitrarily augmentable counter-evidence.

[17] For details cf. Klaua (1969), p. 173.

[18] This is a short notation for the function assigning to each w in the domain the value $\kappa(w \mid A)$.

[19] This was pointed out to me by Godehard Link.

[20] It is easy to link Def. 6 with Gärdenfors (1984). Gärdenfors there discusses contractions and minimal changes of what he calls belief sets, where these belief sets are essentially equivalent to our net contents. Keeping in mind that $\kappa^{-1}(0)$ is the net content of state κ, we may define the minimal change of $\kappa^{-1}(0)$ needed to accept A as $\kappa^{-1}_{A,\alpha}(0)$ for some $\alpha > 0$ (this does not depend on which $\alpha > 0$ we choose). And we may define the contraction of $\kappa^{-1}(0)$ with respect to A as $\kappa^{-1}(0)$, if $\kappa(\bar{A}) = 0$, and as $\kappa^{-1}_{A,0}(0)$, if $\kappa(\bar{A}) > 0$. It is then easy to prove that contractions and minimal changes so defined have all the properties (1)–(21) Gärdenfors (1984, pp. 140–142) wants them to have.

[21] A self-comment: In my (1983), I explicated the notion that A is a reason for B relative to SCFs (which I there called selection functions). This has now turned out to be inadequate, but it is easily repaired: *A is a reason for B* in the state κ iff B is believed in κ with greater firmness given A than given $\bar{A}$, i.e. iff $\kappa(\bar{B}|A) > \kappa(\bar{B}|\bar{A})$ or $\kappa(B|A) < \kappa(B|\bar{A})$. The rest of the paper is easily adapted to this new definition. (Instead of "A is a reason for B in a given epistemic state" one may also say that A *means* B in that state. This is, it seems to me, the most basic meaning of meaning on which other (linguistic) concepts of meaning may be built.)

[22] Cf. e.g. Spohn (1980), Theorem 1(d) and (e). – Indeed, I wonder how far the mathematical analogy could be extended. What I have shown is that the probabilistic theory of dependence, independence, and conditionalization can be carried over to OCFs. The Definition 7 of generalized conditionalization suggests that the concept of a *mixture* may also be meaningfully carried over from probability measures to OCFs. This might be worth exploring. One essential point of dissimilarity is that, as far as I see, there is no meaning to a theory of integration within the theory of OCFs.

[23] Cf. e.g. van Fraassen (1976).

[24] Cf. Harper (1976, pp. 87ff.), or my (1986). The dimensionally well-ordered families of probability measures introduced in the latter paper are the counterparts to our WOPs; and these families represent Popper measures just as WOPs represent SCFs according to Theorem 1.

[25] The idea is essentially due to Kurt Weichselberger. I have merged his idea with an idea I found in Skyrms (1983, p. 158).

[26] One may perhaps conclude that I should have carried through the whole business of OCFs within a nonstandard framework form the start. However, I am happier with the standard version presented, and I did not want to burden my theory with nonstandard number models.

[27] Cf. e.g. Nute (1980, ch. 1 and 3).

[28] Gärdenfors (1981, p. 207), e.g., explicitly accepts this thesis.

[29] Cf. also Levi (1980, p. 7).

[30] On p. 188, Shackle (1969) says that the assumption of the cardinality of his tool is "by no means indispensible to its main purpose".

REFERENCES

Ellis, B. (1979) *Rational Belief Systems*, Oxford, Blackwell.

van Fraassen, B. C. (1976) 'Representation of conditional probabilities', *Journal of Philosophical Logic* **5**, 417–430.

Gärdenfors, P. (1979) 'Conditionals and changes of belief', *Acta Philosophica Fennica* **30**, 381–404.

Gärdenfors, P. (1981) 'An epistemic approach to conditionals', *American Philosophical Quarterly* **18**, 203–211.

Gärdenfors, P. (1984) 'Epistemic importance and minimal changes of belief', *Australian Journal of Philosophy* **62**, 136–157.

Harper, W. L. (1976) 'Rational belief change, Popper functions and counterfactuals', in W. L. Harper, C. A. Hooker (eds.), *Foundations of Probability Theory, Statistical Inference, and Statistical Theories of Science, vol. I*, Dordrecht, Reidel, pp. 73–115.

Jeffrey, R. C. (1965) *The Logic of Decision*, Chicago, University Press (2nd ed. 1983).

Klaua, D. (1969) *Allgemeine Mengenlehre*, Berlin, Akademie-Verlag.

Levi, I. (1980) 'Potential surprise: Its role in inference and decision-making', in L. J. Cohen and M. Hesse (eds.), *Applications of Inductive Logic*, Oxford, Clarendon Press, pp. 1–27.

Levi, I. (1983) 'Truth, fallibility, and the growth of knowledge', in R. S. Cohen, M. W. Wartofsky (eds.), *Language, Logic, and Method*, Dordrecht, Reidel, pp. 153–174.

Lewis, D. (1973) *Counterfactuals*, Oxford, Blackwell.

Lewis, D. (1980) 'A subjectivist's guide to objective chance', in R. C. Jeffrey (ed.), *Studies in Inductive Logic and Probability*, vol. II, Berkeley, University Press, pp. 263–293.

Nute, D. (1980) *Topics in Conditional Logic*, Dordrecht, Reidel.

Pollock, J. L. (1976) *Subjunctive Reasoning*, Dordrecht, Reidel.

Shackle, G. L. S. (1969) *Decision, Order, and Time in Human Affairs*, Cambridge, University Press, 2nd ed.

Skyrms, B. (1983) 'Three ways to give a probability assignment a memory', in J. Earman (ed.), *Testing Scientific Theories. Minnesota Studies in the Philosophy of Science*, vol. X, Minneapolis, University of Minnesota Press, pp. 157–161.

Spohn, W. (1980) 'Stochastic independence, causal independence, and shieldability', *Journal of Philosophical Logic* **9**, 73–99.

Spohn, W. (1983) 'Deterministic and probabilistic reasons and causes', *Erkenntnis* **19**, 371–396.

Spohn, W. (1986) 'The representation of Popper measures', *Topoi* **5**, 69–74.

ARTHUR W. BURKS

THE LOGIC OF EVOLUTION, AND THE REDUCTION OF HOLISTIC-COHERENT SYSTEMS TO HIERARCHICAL-FEEDBACK SYSTEMS*

1. EXAMPLES OF THE TWO KINDS OF SYSTEMS

The distinction between holistic-coherent systems and hierarchical-feedback systems is best explained by means of examples. We will use examples of both natural systems and of man-made systems, such as philosophical systems and computing systems.

An organism is a holistic-coherent system, and rational-intentional-conscious man is an especially interesting and complicated holistic-coherent system. An ecology is holistic-coherent. Biological evolution and cultural evolution are holistic-coherent systems. The non-linear dynamic systems studied by physicists are holistic-coherent. Many natural holistic-coherent systems are very complex and non-linear, and are capable of learning and adaptation.

Many philosophers view reality and knowledge as holistic-coherent systems: Hegelians and advocates of coherence theories of truth are examples. Gestalt phenomena are holistic. Philosophers who say that science is essentially subjective, and relative, or that science evolves by holistic jumps, are treating science as a holistic-coherent system.[1]

A hierarchical-feedback system is composed of atoms or building blocks at each level, with cyclic feedback connections within levels and between levels. On each level the atoms of that level are compounded into units which become the atoms or building blocks of the next higher level.

Computing systems – ranging all the way from simple computers to vast computer networks – are paradigms of hierarchical-feedback systems. Logical switches, memory elements (e.g., flip-flops), wires for communication, input primitives (e.g., keys), and output primitives are atoms or building blocks at a low hardware level. These are compounded into registers, counters, address, communication buses and switches, etc. of the next higher hardware level. In turn, these are compounded into arithmetic units, memories, controls, etc. to make a

William L. Harper and Brian Skyrms (eds.), Causation in Decision, Belief Change, and Statistics, II, 135–191.

computer. Several computers may be networked into a single large system.

The organization of building blocks and the distribution of control at each hardware level of a computer is as important as the properties of the building blocks themselves. In this sense a whole is greater than the sum of its parts. My own view is that human consciousness has to do with organization and control in the human computer.

Computers can also be treated as software systems, again, with many levels. The machine language is on the lowest level. At each level the atoms are instructions and the compounds are programs. Automatic programming languages (e.g., compilers) translate between levels. Within basic limits, software and hardware are logically interchangeable.

Natural systems as described by philosophical materialists and by modern physics and chemistry are hierarchical-feedback systems. The modern genetic and biochemical theory of biological evolution depicts a hierarchical-feedback system. We will use many concepts and principles derived from evolutionary processes.

2. MY PHILOSOPHY OF LOGICAL MECHANISM

I call my philosophy "the philosophy of logical mechanism."[2] This philosophy is a generalization of traditional materialistic and mechanistic views which relies heavily on stored-program computers and robotic extensions of them as models. In my sense of "logical mechanism," automatic devices, computers, robots, natural organisms, genetic programs, learning mechanisms, and evolutionary processes are all *logical mechanisms*.

A pivotal theory of the philosophy of logical mechanism is the "man–robot thesis," namely, for each human, there is (in principle) a computer-robot which can perform the same natural functions as the human. Since man is a holistic-coherent system, the man–robot thesis is a basic reductionist claim. I have advanced a finite type of argument for this claim, based on the existence of a threshold of sensation, and similar considerations. Such a general argument gives no insight into how the internal mechanism works, however. Modern physiology, biology, and psychology give many explanations, and I have offered computer accounts of intentional goal-directedness and functional consciousness.[3]

My finitistic argument for the man–robot thesis can be adapted to apply to many holistic-coherent systems, including finite runs of evolution, both biological and cultural. But again, such a philosophical argument yields no understanding of how the system works internally. By means of a robotic model of evolution I have indicated how intentional goal-directedness and functional consciousness evolved as efficient adaptations to the environment.[4] In the present paper I want to generalize these results and show how hierarchical-feedback systems learn and adapt, by developing a "logic of evolution."

Since the thesis of this paper is that holistic-coherent systems are in principle reducible to hierarchical-feedback systems, something needs to be said about reducibility. I have elsewhere analyzed this concept in terms of one system being an *embedded subsystem* of another, more complicated, underlying system.[5] Consider, for example, the ideal gas law $PV=kT$. It describes a system S_e of macrostates which is an embedded subsystem of an underlying system S_u of particles in microstates. The kinetic theory of gases shows how to reduce S_e to S_u. The ideal gas law is the theory T_e of subsystem S_e. Let T_u be a hypothetical complete description of the underlying system S_u. Note that T_u is too complicated for a human to possess, while T_e is simple and useful.

This example of reduction is exceedingly simple. Thus macro-quantities are obtained by simply averaging micro-quantities. In more typical reductions microproperties are combined in more complicated ways to obtain macroproperties. This is so for most economic, learning, and evolutionary systems.

An adequate computer model of a system represents an embedded subsystem of that system.[6] Since software is reducible to hardware, the existence of a computer model of a system implies the reducibility of that system to a logical-mechanical system. For this reason, computer software for simulating evolutionary processes or a computer based on evolutionary principles would help to show that holistic-coherent systems are reducible to hierarchical-feedback systems. Starting in Section 4 we will use a new type of computing system, called "classifiers," to develop the logic of evolution.

A few words are needed on the relevance of the logic of evolution to the topic of this conference, "Probability and Causation." The philosopher Charles Peirce saw correctly that evolution is a long and continuous process, beginning with an astronomical and physical-chemical phase, proceeding through biological evolution to the development of social groups, and then producing culture, including modern science. Moreover, he saw that probability plays a key role at every step of evolution.

In adapting to nature, an evolving system acquires some kind of fit or match with nature, some kind of correspondence with or adequate models of relevant aspects of nature. Fitness includes operating in accord with the laws of nature, as well as learning about the properties and locations of specific objects and events in the environment. There are probabilistic as well as determinate laws of nature. The logic of evolution can be used to learn about both.

This paper is also relevant to the subject of this conference in that it shows how Bayesean confirmation is a special case of the logic of evolution.

3. LOGIC OF EVOLUTION

My proposed *logic of evolution* has three parts:

(1) Logic of indirect evaluation and selection
(2) Learning by repeated competition
(3) Logic of genetic discovery.

We will explain the first part in the present section. The second and third parts are derived from the learning and discovery algorithms of my colleague John Holland's radically new computing system which he calls "classifiers." Basic classifiers will be described in Sections 4, 5, and 6; the "bucket-brigade" learning algorithm in Sections 7 and 8; and the genetic discovery algorithm in Section 9.

"Learning" is meant here in the sense of inquiry and problem-solving, not in the sense of remembering or memorizing. Thus much of what John Dewey, for example, wrote about logic belongs to the second part of evolutionary logic. Many people have written informally on the logic of discovery, as contrasted to logic of argument, and such material belongs to the third part of evolutionary logic.

The basic principle of the *logic of indirect evaluation and selection* is the evaluation and selection of entities according to the properties of their consequences or productions. We will illustrate this principle by comparing two very different examples: (a) A Bayesean verification process, and (b) a simplified evolutionary example.

Bayesean example. There are two boxes of marbles, with known and different frequencies. One box is chosen at random, and repeated random drawings are made with replacement. Bayes' theorem is applied to evaluate the relative probabilities, and after a suitable number of drawings one hypothesis is chosen as being probably true.[7]

Evolutionary example. There are two organisms, with different genomes, living in a (static) environment. Each organism reproduces parthenogenetically, and the descendents compete for the same resources of the environment. This process is iterated until most surviving organisms are descendents of one of the original organisms. Thus the evolutionary process has selected one genome as better fitted to the environment than the other genome.

I will next give a general characterization of the logic of indirect evaluation and selection, using these very simple examples as illustrations.

An evolutionary system has two or more *producers* operating in a context or *environment*. A producer repeatedly generates *products*. [In the Bayesean example the producers are hypotheses, or possible boxes; the environment consists of the random drawing procedures; and the products are the individual drawings. – In the evolutionary example the producers are the genotypes, the environment is the natural environment, and the products are the organisms which function and reproduce in this environment.]

Producers are relatively stable and long-lived, products are relatively transient, and producers generate a succession of products. Products are *directly evaluated* by the environment. By means of repeated product evaluations, the environment *indirectly evaluates* and ultimately *selects* from among the producers. [In the Bayesean example, each drawing has a probability distribution relative to the hypotheses. By Bayes' theorem for calculating inverse probabilities from direct probabilities, this distribution affects the relative likehood of the hypotheses, and ultimately one of the hypotheses may be selected. – In the evolutionary example, each organism has some degree of fitness to the environment: it can use the materials of the environment to grow, and perhaps reproduce copies of itself. Typically the organisms of the two genotypes will survive and reproduce at different rates; ultimately this will amount to the environment selecting one of the genotypes and rejecting the other.]

As these and other examples show, the logic of indirect evaluation is applicable to many apparently diverse fields. Industries produce products from raw materials or parts, trucks move these products, and merchants distribute them. Machines and tools in the hands of users are "producers" in one sense of the term. Educational institutions generally increase the knowledge and skills of their students. Organisms respond in more or less uniform ways to inputs. Computers transform inputs into outputs.

Thus there are inputs to a producer, which transforms each into an output product. There is an "environment" in the most general sense which supplies the inputs, evaluates the outputs, and selects from among the producers.

The logic of indirect evaluation and selection may operate at many levels in the same system. Thus machines and workers may be tested and evaluated by a unit of an organization, that unit may be monitored by a section, the section by a division, and the whole organization competes with other organizations of like kinds. Within a human, bodies and antibodies compete, and so do desires, emotions, and purposes.

It is customary to limit the term "logic" to symbolic processes. But, following Hegel and Peirce, I will call a natural process "logical" if it follows some logical law. As indicated above, economic and evolutionary processes have many logical aspects. It is reasonable to broaden the scope of logic this way, since symbolic processes evolved gradually from non-symbolic processes.

4. BASIC CLASSIFIER SYSTEMS

Philosophical antecedents. A "classifier" is a formal conditional or hypothetical, used as an instruction in a new type of electronic computer. Classifiers have historical antecedents in philosophy.

Bishop Berkeley analyzed the external world into conditional laws connecting sense experiences. Following Berkeley, Charles Peirce analyzed man's knowledge of the environment into "practical conditionals" and man's reasoning and reacting processes into "habits." A practical conditional is a hypothetical statement or rule, in one of two forms. In one form, the antecedent describes an experiment or action and the consequent describes an observable result. In the other form, the antecedent describes a goal and some circumstances, and the consequent describes an action for achieving that goal in those circumstances. Both singular object statements and general law-like statements are to be analyzed into practical conditionals; probabilities may be involved. A habit, for Peirce, is an embodiment (consciously or unconsciously) of a practical conditional in thought or action.

The stimulus-response reactions which behaviorists take to be basic can be expressed as conditionals. The conditioned reflex discovered by Ivan Pavlov is a stimulus-response connection that develops out of a natural reflex by repeated association. This development involves the sequence of hypotheticals shown in Figure 1.

Sequence of hypotheticals involved in the development of a conditioned reflex (after Pavlov).

(1)	IF sight-of-food	THEN salivary-secretion	[Initially given natural reflex]
(2)	IF sight-of-food & sound-of-bell	THEN salivary-secretion	[Developed by repeated association]
(3)	IF sound-of-bell	THEN salivary-secretion	[Resultant conditioned reflex]

Fig. 1. Development of a conditioned reflex.

I think that Peirce's theory of "habits" (mental or behavioral conditionals) and their evolution into cooperating "bundles of habits" constitutes a good rough account of how the mind reasons and learns. According to Peirce, single habits develop and strengthen, form clusters, and combine with other habits to evolve into bundles of habits. The underlying law is the psychological law of association. From the internal side this process is guided by the goals of the system. On the external side, some habits fit well with the environment and are maintained, other habits are not useful in the environment and are rejected or modified. Peirce said that this learning process is closely analogous to, indeed is a special case of, biological evolution.

Peirce did not give any details on how this kind of system operates, nor could he have done so, given the absence of knowledge about genetics and computers in his day. Nevertheless, I think a radically new type of computing system called "classifiers" is a good computer formalization of what Peirce aimed at. Classifier systems and associated algorithms for learning and discovery have been developed by my colleague John Holland, and I have been developing novel computer architectures for the rapid computation of classifier programs.

This research has been carried out in our Logic of Computers Group at The University of Michigan, a faculty-student research group headed by myself and Holland. The ideas have been validated by simulations, designs, and theoretical results. Classifier programs can, of course, be run on standard machines, but we can run them much faster when we have special classifier machines.[8]

Basic classifier systems. A classifier system has two kinds of basic entities, *classifiers* and *messages*. A classifier is a hypothetical statement, with one or more *conditions* as antecedents and a computing or action term or *consequent*. Classifiers usually have two conditions, but for the purpose of explaining how conditions and consequents work it suffices to consider only the one-condition case. If a message satisfies the condition of a classifier, the consequent produces a new message from itself and the satisfying message.

Classifier systems will be presented in three stages: basic classifier systems (this section), classifiers with the bucket-brigade learning algorithm (secs. 7 and 8), and classifiers with both the bucket-brigade algorithm and the genetic discovery algorithm (sec. 9).

In a *basic classifier machine* the set of classifiers remains invariant throughout the computation. Computation begins with an initial set of messages. At each main step or *major cycle* of computation, every classifier is applied to all messages to produce a new set of messages; this new set of messages then replaces the old set.

(When the bucket-brigade algorithm is added the message memory will be restricted in size so that only some messages will be carried over to the next major cycle. When the genetic algorithm is added new classifiers will be formed from old classifiers by genetic-like operators so that the set of classifiers will change from major cycle to major cycle.)

There is an obvious analogy between a classifier program and a standard computer program, classifiers playing the role of instructions and messages playing the role of data. But there are two strong disanalogies. In a standard computer program, instructions are linked to data by addresses and in each processor instructions are executed in sequence. In contrast, a basic classifier program has no addresses, since every classifier is applied to every message on the current message list, and hence in each major cycle all classifiers can be executed simultaneously, or in any order. For this reason a classifier computer can be highly parallel.

Messages are binary words (based on the alphabet 0,1). They express inputs to the system, results of calculations that are to be preserved, and outputs from the system. Classifiers are ternary words, based on the alphabet (0,1,#). The number or blank space sign "#" functions somewhat differently in conditions and in consequents. It means "don't care" in a condition. But it means "pass through," or "place the corresponding bit of the satisfying message into the generated message," in the consequent.

Messages are specific, conditions and consequents are general. Consider these two stimulus-response classifiers

(a) If the light is red, then stop.
(b) If the light is green, then go.

For simplicity, assume that only two shades of color can be sensed by the input devices. Figure 2 shows how the messages and conditions can be coded so that conditions can recognize sets of messages. In this example the first bit of a message expresses the generic color (red or green) and the second bit specifies a shade within that color.

	Colors		
Messages	Specific	Generic	Conditions
10	Crimson	Red	1#
11	Burgundy		
00	Yellowish green	Green	0#
01	Bluish green		

Fig. 2. Coding of messages and of conditions.

When the number sign "#" occurs in a condition it means "don't care," so that the message "10" (crimson) and "11" (burgundy) both satisfy the condition "1#" (red). Hence the antecedents of the two classifiers should be formalized

(a′) If 1# then STOP.
(b′) If 0# then GO.

Either shade of red will cause the first classifier to generate a STOP message, and either shade of green will cause the second classifier to generate a GO message.

It is easy to code the output messages so that they produce the appropriate actions; let a first bit "0" in an output message mean "stop" and let "1" mean "go." But it is also useful for a generated message to contain some information from the satisfying message which triggered its production. To accomplish this we stipulate that an occurrence of the number symbol "#" in a consequent means "pass the corresponding bit of the satisfying message into that position of the generated message." In the present case the second bit of an input message specifies the shade, and the generic color (red or green) is implicit in the action (stop or go, respectively), so the machine will pass the second bit of the input message through to the output message. These two operations are accomplished by the classifiers

(α) If 1# then 0#.
(β) If 0# then 1#.

For example, the message 10 (crimson) satisfies the first classifier, which produces the message 00 (stop, crimson), as in Figure 3.

These classifiers are adequate for showing how individual classifiers work. But they are atypical in that they produce output messages directly from input messages. In most situations inputs influence out-

Consider these two classifiers and the message: "10":

First classifier:	If 1# then 0#	(If the light is red, then stop)
Second classifier:	If 0# then 1#	(If the light is green, then go)
Message:	10	(The light is crimson)

This message only satisfies the first classifier, which produces a message 00 as follows.

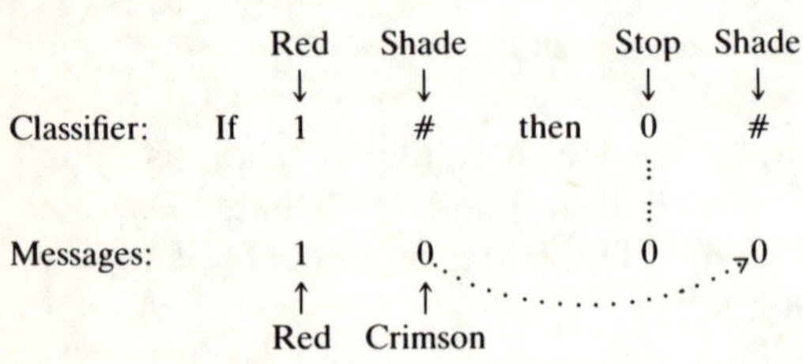

Fig. 3. How classifiers produce new messages.

puts through chains (often long chains) of classifiers linked by internal messages. Moreover, it is in these chains that most learning takes place. This fact has generally been unrecognized or underemphasized by behaviorists.

Logical structure of a basic classifier machine. Figure 4 illustrates successive major cycles of a simple classifier computation. There are two classifiers $\mathscr{C}_1$ and $\mathscr{C}_2$, and two initial messages 0100 and 1011. The antecedent of $\mathscr{C}_1$ is satisfied by 0100, so the consequent of $\mathscr{C}_1$ produces 1000. Similarly, $\mathscr{C}_2$ transforms 1011 into 0111. Thus the first major cycle produces the message set (1000,0111). In the second major cycle, $\mathscr{C}_2$ transforms 1000 back into 0100, and $\mathscr{C}_1$ transforms 0111 back into 1011, yielding the new message set (0100,1011). This was the initial message set, so in this simple system the two message sets merely alternate from one major cycle to the next.

	Classifier set:	Initial message set:	Alternate message set:
$\mathscr{C}_1$:	0 1 # # ⊃ 1 0 # #	0 1 0 0	1 0 0 0
$\mathscr{C}_2$:	1 0 # # ⊃ 0 1 # #	1 0 1 1	0 1 1 1

Initial	First	Second	Third	
0 1 0 0 ------	$\mathscr{C}_1 \rightarrow$ 1 0 0 0 ------	$\mathscr{C}_2 \rightarrow$ 0 1 0 0 ------	$\mathscr{C}_1 \rightarrow$ 1 0 0 0	------
1 0 1 1 ------	$\mathscr{C}_2 \rightarrow$ 0 1 1 1 ------	$\mathscr{C}_1 \rightarrow$ 1 0 1 1 ------	$\mathscr{C}_2 \rightarrow$ 0 1 1 1	------

Major cycles

Fig. 4. Major cycles of a basic classifier computation.

This simple picture needs to be enlarged in several ways. There are generally many more messages than classifiers, and classifiers usually have two conditions. A typical small classifier system has about 32 messages of 16 bits each, and 1024 classifiers of two conditions. Input and output messages need to be distinguished from internal messages, input messages coming from sensors or other input devices and output messages going to actors or other output devices. Marker bits in a preassigned field can be used to distinguish different categories of messages. Classifiers can both recognize and produce such markers, tags, or labels.

In a two-condition classifier the second condition normally plays a controlling or trigger role: the first condition and consequent only function in the way we have explained *if* the second condition is satified by at least one message. The second condition may be negative. In that case, the first condition and consequent function to make a new message only if *no* message satisfies the second condition.

The essential recursion executed by a basic classifier system is a repetitive set operation. At each major cycle the set of classifiers is applied to the old message set to yield a new message set, which becomes the old message set for the next major cycle. (Also, input messages need to be added to the message list and output messages derived from it.) But while it is logically possible to process all classifiers against all messages in parallel, it is not feasible to build such a large logical switch with present technology. (One can speculate on novel methods, e.g., the use of holography.)

We therefore propose processing the classifiers sequentially, defining a *minor cycle* as the time taken to process one classifier against all messages. A major cycle will then consist of as many minor cycles as there are classifiers, plus time for adding input messages, sending messages to output devices, and replacing the old message set by the new message set.

Figure 5 shows how a basic classifier machine is organized. Since there are many more classifiers than messages, and classifiers are much longer than messages, the classifier store is very much larger than either message store.

We will analyze the operation of Figure 5 for typical major and minor cycles. At the beginning of a major cycle all classifiers are in the classifier store, there is a set of messages in the old message store, and the new message store is empty. Fresh messages from sensors or other input devices are added to the old message store. Then the machine processes each classifier in turn, but in any order, the processing of a classifier occupying one minor cycle.

At the beginning of each minor cycle a classifier moves from the classifier store to the classifier processor. The parallel comparator switch can compare one condition against all of the messages in the old message store simultaneously. The second condition of the classifier is compared with the old messages first. If the second condition is satisfied, then the first condition is compared. For each message satisfying it (if any), the classifier generates a new message and places it in the new message store. At the end of the minor cycle the classifier moves back to the classifier store.

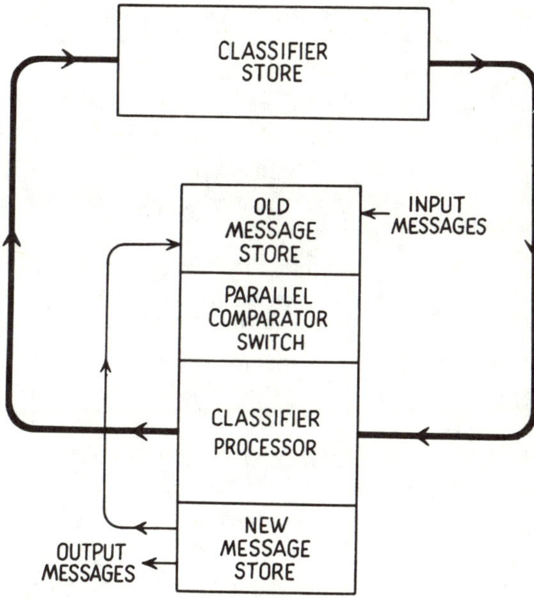

Fig. 5. Basic classifier machine.

At the end of the major cycle the old messages are erased and replaced by the new messages, and messages are transmitted to actors or other output devices.

We will conclude with some remarks about parallel computation. The parallel comparator switch and old message store are often called an "associative memory" or "content-addressable memory." Psychological association does involve content comparison, though this could be done serially rather than in parallel. Moreover, psychological association involves other factors, as in the development of a conditioned reflex (see Fig. 1 above). In any case, a parallel processor that compares a classifier condition with all old messages simultaneously is very much faster than a serial processor.

5. SPECIFIC AND GENERIC UNIVERSALS

We can motivate the use of a more inclusive alphabet for classifiers (0,1,#) than for messages (0,1), by means of the distinction between general properties and specific properties. In philosophical terminology, this is the distinction between generic universals and specific universals. Conditions and consequents in classifiers represent general properties or universals, while messages represent specific properties or universals. Some generalization and classification may take place in the input equipment of a classifier machine, but as far as the machine itself is concerned, a message expresses a quite specific universal or property.

This distinction between specific messages on the one hand and generic conditions and consequents on the other hand, has some analogy to Peirce's distinction between First and Third as psychological categories: for Peirce a First is a specific experience; while a Third is a general concept, embracing a class of specific experiences which the mind treats as a unit.

A Zen Buddhist once said "Since all can be sensed, there is no need to speak." but this is clearly wrong, for only the specific can be sensed, while life and thought depend on the use of generals (as Peirce emphasized). Organisms respond to objects in terms of their general properties, and the laws of nature connect general properties to one another. The very form of language is general, for the same symbol can be spoken or written in a wide variety of forms or patterns. This is Peirce's distinction between *types* and *tokens*. Likewise, the concepts expressed by language are general.

This Zen saying is also refuted by the difficulties computer scientists have had with the pattern recognition problem. Consider the human ability of reading handwritten letters of the alphabet. It is relatively easy to build sensing devices that can code specific letter tokens into messages, so *if* the Zen saying *were* true, it *would be* easy to build a machine that could recognize handwritten alphabetic characters. But computer scientists have found that it is very hard to build such a machine. They have worked on the pattern recognition problem for a third of a century, and yet no one knows how to build a machine that can read handwritten alphabets. The reason for this is that handwritten alphabetic characters are generic patterns which are very complex sets of specific patterns.

The essential difficulty in recognizing handwritten characters, and in many pattern recognition problems, involves the interplay of gradual variation and context. See Figure 6, "The Vagueness Continuum from *A* to *R*." This continuum begins with a clear-cut letter *A* and proceeds

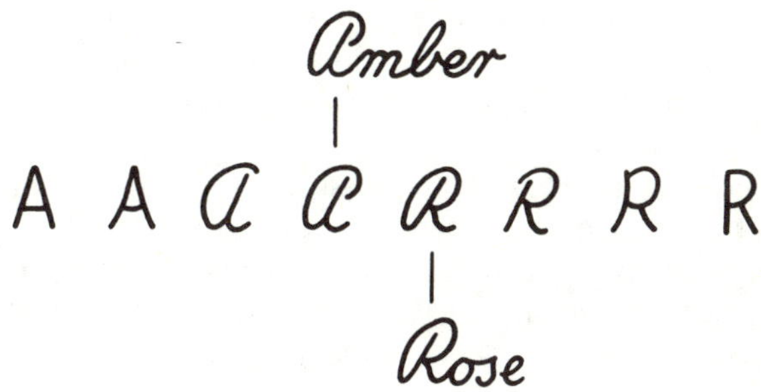

Fig. 6. The vagueness continuum from *A* to *R*.

by small increments to a clear-cut letter *R*. When a textual character is of one of the intermediate forms of Figure 6, resolution of the ambiguity requires looking at the context and ascertaining the linguistic and subject-matter constraints which the context imposes. In difficult cases the context is also ambiguous, so the interpreting system needs to go back and forth between letters and context to resolve the ambiguities. Such ambiguous texts constitute small holistic-coherent systems.

Classifiers are so-called because they classify messages into subsets. Derivatively, basic classifier programs can classify sequences of input messages into subsets, and hence can be used for pattern recognition. When the learning and discovery algorithms are added to basic classifiers, the extended system constitutes a powerful new tool for solving pattern recognition problems.[9]

We will explain these extended classifier systems below (secs. 7, 8, and 9), but here is a good place to point out an important difference between the operation of a basic classifier system and the operation of one of these extended classifier systems. A basic classifier program does not change during its operation, so that for a given initial message list and sequence of input messages it always responds with the same sequence of output messages. This is no longer true when the learning and discovery algorithms are added.

The bucket-brigade learning algorithm restricts the message set by an evaluation and selection process, so that only some of the messages produced by classifiers during a major cycle are carried over to the next major cycle, those messages not selected being destroyed. To get their messages carried over, classifiers must compete in a free market exchange or auction. Each classifier has a *strength* which functions as economic capital: a classifier pays out some of its strength to get its message carried over to the next major cycle, and receives a payment of strength from any classifier which uses this message successfully during the next major cycle.

Under this competitive system, classifiers will differ in their success in getting the messages they produce carried over and used. Consequently, classifier strengths in a bucket-brigade program will generally vary over repeated runs made with exactly the same sequence of input messages. As classifier strengths change, so will the internal and output streams of messages. Hence, for a bucket-brigade program, the output stream of messages will generally vary from run to run even with a repetition of exactly the same input message sequence.

This will still be so when the genetic algorithm is added to a bucket-brigade classifier program. In this extended system classifier strengths are also used to select from among classifiers. Weak or "poor" classifiers are eliminated and strong or "wealthy" classifiers are used for the genetic production of new classifiers. These new classifiers replace the eliminated classifiers, and the revised set of classifiers is then tested by the bucket-brigade algorithm. New classifiers will usually contain new conditions or consequents, so the genetic algorithm creates new concepts as well as new rules.

To conclude this discussion of generality and specificity we note that the ternary alphabet (0,1,#) yields very simple measures of generality and specificity with respect to the binary alphabet (0,1). The generality of a condition increases with the relative frequency of don't-cares (#'s) and the specificity decreases as the generality increases. For example, the completely general condition #### recognizes all sixteen messages of length four, the less general condition 0### is satisfied by eight messages, the more specific condition 01## recognizes four messages, and the completely specific condition 1010 is satisfied by only one message, namely 1010. A measure of specificity is used in the bucket-brigade algorithm to help determine how much a classifier will bid to get its message included in the message set for the next major cycle.

6. TRANSITION TO LEARNING

A robot problem. Many simulations have been made of classifier systems solving problems with the learning and discovery algorithms.

Examples include control of an oil pipeline distribution system, learning to play poker, learning to distinguish between alphabets, and a robot locating an object in an environment.[10] We will describe a task of the latter type and then show how the bucket-brigade learning algorithm would learn to do this task.

The following problem occurred on an intelligence test I took as a young boy. I was shown a circle with a gap in it, and told that the circle represented a fence around a field and the gap represented the entrance to the field. A baseball was lost somewhere in the field and I was to find it. I was then given a pencil and asked to trace the path I would follow in my search. A good answer was a systematic scan, such as spiral from outside in; a random walk was a poor answer.

Imagine now a robot in a large room containing objects of various kinds in different places. The robot has a television camera with which it can see an object nearby which is in the field of vision of the camera. The robot can rotate its camera and move back and forth around the floor, and it can distinguish the different kinds of objects in the room and recognize the wall of the room. Figure 7 is a block diagram of the robot.

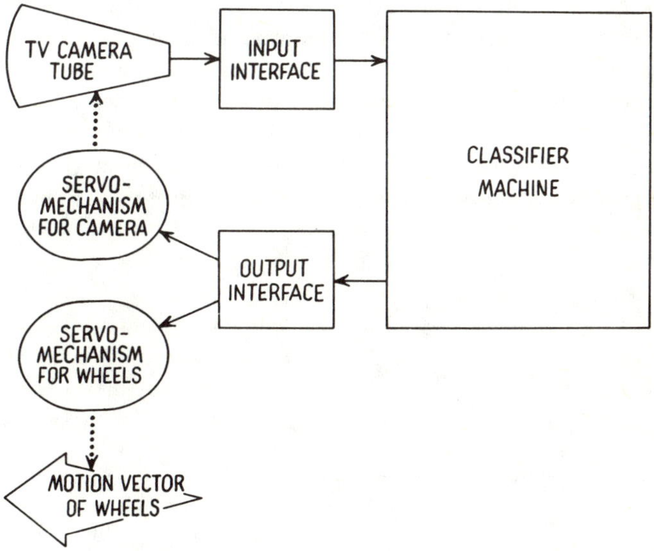

Fig. 7. Classifier-based robot.

With this background we can formulate an *object-location problem* somewhat analogous to the lost baseball problem. The robot is placed somewhere in one end of the room, and it is to move around and find an object of a specified kind placed somewhere in the other end of the room. When the robot locates this object it should position itself next to the object, face it, and stop. Note that built into the robot is the capacity to recognize when it reaches its goal and the problem is solved.

As in the baseball problem, a solution to the object location problem will be a general skill or bundle of habits which enables the robot to solve efficiently many specific cases that vary in detail. Consider now the basic classifier systems described in Section 4. It follows from the finite character of the object location problem that there is a basic classifier program which will enable a robot with the appropriate sensory and action equipment to solve it. That is, there exists a set of basic classifiers Γ which will direct the robot to respond to its inputs so as to search for and find the designated object in each case.

Let us look at the structure of the execution of this classifier program Γ. Figure 8 shows a small segment of history. The classifiers of Γ are labelled $\mathscr{C}_1$, $\mathscr{C}_2$, $\mathscr{C}_3$, etc. and the messages produced are labelled M_1, M_2, M_3, etc. Input and output messages are placed in circles. Note the series-parallel structure of the computation, with input messages entering and output messages exiting. The actual inputs and hence the actual sequence of classifier uses and outputs, will vary from run to run.

Series-parallel conditionals. Built into the robot is the ability to recognize when it has reached its goal. Suppose in Figure 8 that

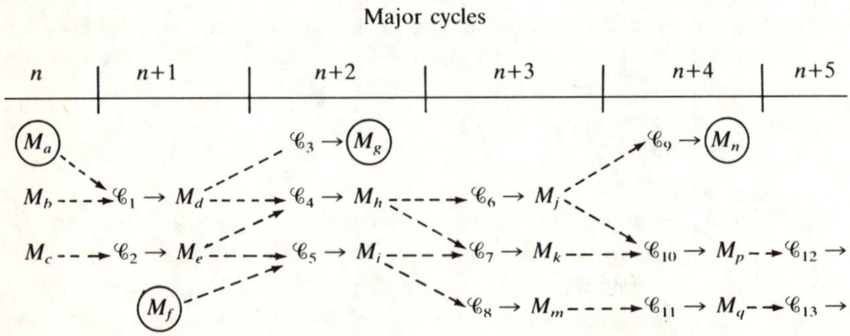

Fig. 8. History segment of a basic classifier program.

message M_n, which is an output message, causes the robot to take the last step to the goal. Since the classifier program Γ consists of conditionals operating in series and in parallel, it is reasonable to view this program as an extended *series-parallel conditional.*

This concept of a series-parallel conditional is a natural extension of the logical concept of a conditional. For example, if the three conditionals

If *Px*	then	*PxQx*
If *PxQx*	then	*PxRx*
If *PxRx*	then	*PxSx*

are applied to *Pa* iteratively the result will be *PaSa.*

A basic classifier system is deterministic, and its sets of input, internal, and output states are finite. Hence it is equivalent to a finite automation. It follows that for each input sequence a basic classifier system will produce a unique sequence of outputs.

One could change the determinate conditional connection of a classifier to a probabilistic connection, so that when the conditions of a classifier are satisfied it produces a probabilistic distribution of messages, that is, different messages with various probabilities. The classifier system would then be equivalent to a probabilistic automation, and the series-parallel conditional defined by the classifier system would be a generalized Markov chain, that is, a parallel interacting Markov chain with inputs and outputs.

However, we add probabilities to classifiers in a more fundamental way. Under the bucket-brigade algorithm each classifier will have a strength, and these strengths will play an important role in a probabilistic process for selecting a subset of generated messages to be carried over to the next major cycle. When the genetic algorithm is added the set of classifiers will change from major cycle to major cycle by a probabilistic genetic process.

7. LEARNING BY REPEATED COMPETITION

Application of the logic of the indirect evaluation and selection to learning. We are now in a position to formulate the learning problem for classifier systems. Consider two different basic classifier programs Γ and Γ′, each of which solves the object location problem with some degree of efficiency. We can apply both Γ and Γ′ to successive instances of the object location problem, score each on its performance, and ultimately choose one of the classifier programs as superior to the other.

This procedure is clearly an application of what I called the logic of indirect evaluation and selection. Each classifier program is a producer operating in the room with objects. A classifier program converts a sequence of environmental stimuli into a sequence of output actions as product. Each product is directly evaluated according to its success in the environment. The producers are indirectly evaluated by repeated production evaluations and ultimately one is chosen as superior to the other.

This is an application of the logic of indirect evaluation and selection to the whole classifier system, a "global" application. But this logic can also be applied "locally," within the system. For a classifier system is a system of classifiers, each of which is a producer converting raw materials (old messages) into products (new messages). Hence the logic of indirect evaluation and selection can be applied to the component classifiers of a classifier system. This is, in fact, how bucket-brigade learning works. A classifier system with bucket brigade evaluates each of its classifiers by repeated tests and uses only the best of them, gradually learning from its successes and improving its performance.

In this local application of the logic of indirect evaluation and selection, the parameter of evaluation is called the "strength" of a classifier. The bucket-brigade algorithm introduces competition between classifiers in making their messages available to other classifiers. The strength of a classifier is at once a measure of the past success of a classifier in producing messages that are used and the main constituent in the system's decision as to which classifier message will be used. The strength of a classifier functions somewhat like the capital of a producer in a market economy, and we will sometimes take advantage of this analogy by using economic terminology in describing the operation of the bucket-brigade algorithm.

The bucket-brigade learning algorithm. A basic classifier system is like an inefficient planned economy in which products are produced by quota independently of demand and quality. For in a basic classifier system every message produced in one major cycle is carried over to the next major cycle, independent of the utility of that message. In contrast, a classifier system with bucket brigade uses a competitive market to select a subset of the generated messages for use in the next major cycle. This competitive market involves limiting the size of the message list, using an auction to decide which messages are preserved, and making payments according to the outcome of the auction.

The bucket-brigade learning algorithm begins with an initial classifier program, consisting of a set of classifiers each with its assigned strength. In writing this program the programmer should include all classifiers that he or she thinks might contribute to solving the problem, since the bucket-brigade algorithm will evaluate all classifiers and render ineffective those that are useless. Initial strengths may be assigned arbitrarily, or they may be made to reflect the programmer's guess as to the importance of each classifier.

The robot is then directed to make problem solving runs. If the initial set of classifiers is sufficient, the robot will solve the problem by trial and error, perhaps very inefficiently. (The case of a bucket-brigade program which cannot solve the given problem will be discussed in Section 9, "Logic of Genetic Discovery.") When it solves its problem the robot rewards itself in a manner to be explained later in this section. As the robot repeatedly makes problem-solving runs and rewards itself, these rewards will gradually feed back to strengthen those classifiers which contribute most to these successes, until the robot has a classifier program that is efficient for that problem.

In any classifier system a classifier produces an output message from a message satisfying its first condition when one or more messages satisfy its second condition. Thus even in a basic classifier system a classifier resembles a producer in an economic system, such as a farm, mine, smelter, manufacturer, builder, etc. Under the bucket-brigade algorithm this economic analogy is greatly strengthened because a classifier risks some of its associated strength in an auction where it competes with other classifiers to get its message put on the (now limited) message list that is carried over to the next major cycle.

The bucket-brigade auction operates as follows. When a classifier produces a message it makes a bid to get this message sent to the next major cycle. The amount bid by the classifier is some function of the strength of the classifier, the specificity of the classifier's conditions, a fractional constant, and perhaps other factors. This function might be the product of these factors. The constant is set so as to limit the fraction of its strength that the classifier risks in making a bid. A specific classifier can afford to bid more than a competing general classifier, since the specific classifier contains more detailed information about the situation, and hence is more likely to have its message used and paid for at the next major cycle than is the case for the general classifier.

The bucket-brigade algorithm modifies the bids probabilistically and then selects those messages with the highest modified bids to constitute the message set for the next major cycle. Probabilistic variations are introduced so that messages with low bids, which might prove to be useful nevertheless, will have some chance of surviving, and also to prevent the bucket-brigade algorithm from getting stuck in a rut.

Let us call a classifier which produces a message that is carried over to the next major cycle a "producer," and let us call the classifiers which supplied the messages satisfying the conditions of the producer "the suppliers" of that producer. The bucket-brigade algorithm transfers the bid amount of a producer to its suppliers. The procedure for transferring payments and the timing is given in Figure 9. A payment reduces the strength of the payer and increases the strength of the payee.

Connecting evaluations of the whole to evaluations of the parts. A classifier program is a whole composed of classifiers as parts. Both are

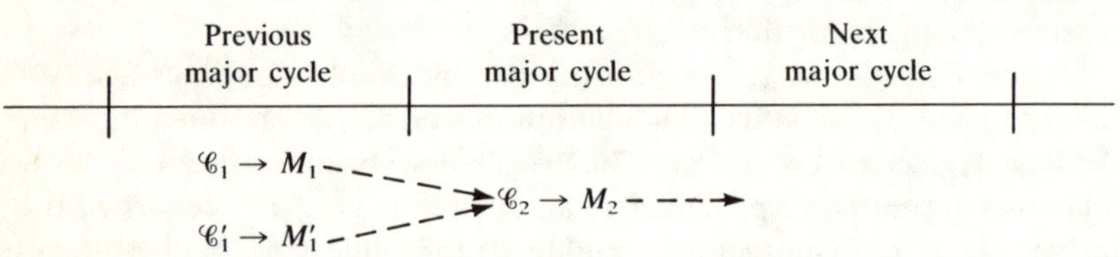

In the previous major cycle classifier $\mathcal{C}_1$ produced M_1 and bid $\mathcal{B}(M_1)$, and $\mathcal{C}'_1$ produced M'_1 and bid $\mathcal{B}(M'_1)$. These bids were sufficient so that messages M_1 and M'_1 were carried over to the present major cycle.

In the present major cycle

(a) Classifiers $\mathcal{C}_1$ and $\mathcal{C}'_1$ are charged for having their messages carried into the present major cycle. Classifiers $\mathcal{C}_1$ is charged the amount $\mathcal{B}(M_1)$, this payment being divided equally among the suppliers of $\mathcal{C}_1$, that is, the producers of messages which satisfied $\mathcal{C}_1$ in the previous major cycle. Corresponding transfers are made for $\mathcal{C}'_1$.

(b) The two conditions of $\mathcal{C}_2$ are satisfied by M_1 and M'_1, so $\mathcal{C}_2$ produces M_2. Then $\mathcal{C}_2$ bids $\mathcal{B}(M_2)$ to have M_2 carried into the next major cycle. This bid is successful so that $\mathcal{C}_2$ now owes $\mathcal{C}_1$ and $\mathcal{C}'_1$ each the amount $0.5\mathcal{B}(M_2)$.

In the next major cycle the amount $\mathcal{B}(M_2)$ will be subtracted from $\mathcal{C}_2$'s strength and the strengths of $\mathcal{C}_1$ and $\mathcal{C}'_1$ will each be increased by $0.5\mathcal{B}(M_2)$.

Fig. 9. Payment procedure of the bucket brigade.

evaluated indirectly, by their outcomes. A program is evaluated in terms of its success in solving a given type of problem. An individual classifier is evaluated by its strength, which is the main factor in deciding its success in getting its messages used by other classifiers. We now need to explain how these two kinds of evaluations are linked, how the evaluations of the whole feed back to the evaluations of the parts.

Consider the strength parameter of a classifier first. Each time a classifier gets its message carried over to the next major cycle its strength is debited for this. If that message is used by one or more classifiers during the next major cycle, the strength of the originating classifier is later credited for those uses. For each such action the classifier makes a profit or takes a loss.

Over repeated problem-solving runs, a classifier's success in producing messages that are used by other classifiers to produce further messages will be reflected in its strength. If its strength increases it will tend to win more competitions to have its messages used, while if its strength decreases it will have the opposite tendency. Thus there is a repeated feedback cycle between strength as a measure of past successes and strength as the main determinant of future successes.

This explains how the bucket-brigade algorithm applies the logic of indirect evaluation and selection to individual classifiers. Evaluation is accomplished by means of classifier strengths. Selection takes the form of more or less use. A strong classifier will be used repeatedly, while a very weak classifier will be used at most rarely. Later, when the genetic algorithm is added, selection will take the form of eliminating weak classifiers and replacing them with offspring of strong classifiers.

We consider next how the evaluations of a whole classifier program enter the program and influence the evaluations of individual classifiers. The robot has the sensory and intellectual capacity to recognize when it achieves its goal. Each time it does so the bucket-brigade algorithm injects an external reward into the system by paying those classifiers which were most directly and immediately associated with the robot's success. There are various options here. We will assume that the bucket-brigade algorithm rewards all classifiers that had winning messages during the final major cycle of the problem run.

Over repeated successful trials these rewards will gradually feed back to enrich other classifiers that contribute messages during a problem-solving run. We will illustrate this process by means of the earlier Figure 8 (History Segment of a Basic Classifier Program). Assume that message M_n causes the robot to take the final step to its goal. Classifiers $\mathscr{C}_9$, $\mathscr{C}_{10}$, and $\mathscr{C}_{11}$ will then receive payoffs. During the next problem-solving run these classifiers will be richer (stronger) and as a consequence will bid more to get their messages carried over. These increased bids will benefit classifiers $\mathscr{C}_6$, $\mathscr{C}_7$, and $\mathscr{C}_8$. As this process is iterated, classifiers which produce messages that are used early in successful runs will gradually profit from the rewards given at the end of each successful run.

In a basic classifier system there is a forward recursion of message generation. The bucket-brigade algorithm adds to this a backward recursion of payments. For its success in the *present* major cycle a classifier pays another classifier for its success during the *previous* major cycle, the payment being made during the *next* major cycle. (See Fig. 9 for an example.) This backwards recursion of payments is fueled by the rewards given to those classifiers which are successful at the time the classifier program reaches its goal.

This explains how the bucket-brigade algorithm connects the evaluations made of the whole classifier program to the evaluations of the individual classifiers which constitute the part of this program.

In medieval times people fought a fire by forming a human line from the nearest water source to the burning building, and passing buckets back and forth. Each person handed a bucket of water to his neighbor and received an empty bucket in return. The person at the head of the line filled the buckets, while the person at the foot of the line emptied the buckets onto the fire. This system was called a "bucket brigade." Hence the name "bucket-brigade algorithm."

We will conclude this section with some general remarks about the problem of measuring the contribution of each part of a system to the success or value of the whole system. In computer science it is called the "allocation of credit" problem. But it appears in many other domains. Thus the problem of determining the contributions of individuals to the success of an organization is a fundamental problem in social organization theory.

There is an analog in personal utility theory, when a person chooses repeatedly from among acts in conditions of uncertainty. Each act is a structured compound of consequences and uncertain statements. The "allocation of credit" problem here takes the form of passing from a person's utility ordering of acts (as reflected in his choices) to his assignment of utilities to consequences and his assignment of subjective utilities to statements. Axiomatic choice theory gives the conditions under which this can be done.[11]

This example of determining the contribution of each "part" of an act to the utility of an act deals with static systems. In learning, adaptation, evolution, and discovery the systems are dynamic. Consider systems of logical atoms such as strings of genes (genomes) and sets of classifiers (classifier programs). Each system generates products: organisms, in the one case and performances in the other. These products are evaluated by an "environment" in a general sense. The bucket-brigade algorithm solves the allocation of credit problem for classifier programs. Evolution solves the allocation of credit problem for genomes by natural selection procedures which produce variable numbers of copies of each kind of gene in the gene pool.

Though these different problems vary widely, in each case a system is studied from the outside to learn about its internal composition, or is influenced from the outside to change its internal composition. Behavioral studies constitute another example. Here one observes the reactions of a finite automaton to various input sequences and from the response sequences infers an internal composition that will produce the observed behavior. There are interesting problems in all these cases concerning the number of trial runs needed to accomplish the result. In general the number of runs needed increases with the complexity of the system and, where the environment influences the system, with the complexity of the environment as well.

8. CLASSIFIER MACHINE WITH BUCKET-BRIGADE LEARNING

Logical structure of machine. The bucket-brigade classifier machine of Figure 10 is an extension of the basic classifier machine of Figure 5, and is best understood in comparison with this earlier figure. There are two interrelated kinds of changes. First, the message set is smaller, to introduce competition. Second, an auction is conducted to decide which generated messages are preserved, and for this purpose both classifiers and messages carry associated information with them. Hence the sizes of both memory units and processing units are different in a bucket-brigade machine from those in a basic classifier machine. These size changes are depicted in the figure.

The classifier store of Figure 10 holds several items along with each

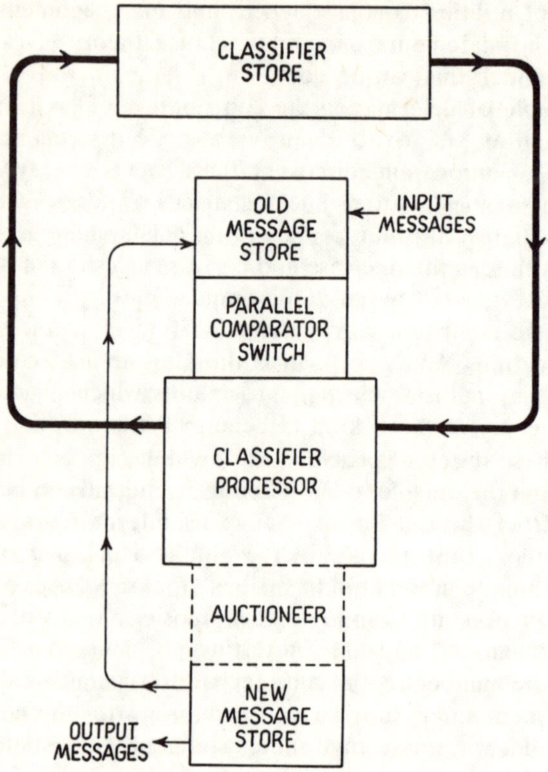

Fig. 10. Classifier machine with bucket-brigade.

classifier: an address, identification number, or "tag," to be used in the accounting process; the present strength of the classifiers; and possibly other items. The generality of the classifier (the relative frequency of "don't care" symbols "#" in its conditions) can be stored with it or calculated each time it is needed.

Since the message set is smaller now than before, both the old and the new message stores are "shorter" in a bucket-brigade machine than in a basic classifier machine. On the other hand, each is "wider" than before, because both old and new messages carry associated items with them. An old message has with it the tag of the classifier which produced it during the previous major cycle. A new message M has with it the bid $\mathcal{B}(M)$, the tag of the classifier $\mathcal{C}$ which successfully made this bid, and the tag of the classifier which supplied $\mathscr{C}$ with an input message.

The classifier processor of a bucket-brigade computer is larger than that of a basic machine because the processor now has two additional functions to perform. First, it must transfer bid amounts from producing classifiers to supplying classifiers. Second, when the processor produces a new message M it now calculates the bid $\mathscr{B}(M)$, and then sends M, its bid $\mathscr{B}(M)$, and the tags of the supplying and producing classifiers to the auctioneer.

When the auctioneer receives this information it compares it with the contents of the new message store. If the new message store is not yet full the auctioneer enters the new information into it automatically. Otherwise the auctioneer compares $\mathscr{B}(M)$ with the smallest bid $\mathscr{B}(M')$ in the new message store. If $\mathscr{B}(M)$ is smaller than $\mathscr{B}(M')$, then message M and its associated information are deleted; if $\mathscr{B}(M)$ is larger than (M'), then message M' and its associated information are deleted and replaced by message M and its associated information.

The transfers of information at the end of a major cycle are more complicated than before. Look again at Figure 9, "Payment procedure of the bucket brigade." During the present major cycle classifier $\mathscr{C}_2$ processes M_1 into M_2 and bids $\mathscr{B}(M_2)$ to get M_2 on the new message list. This bid is successful: it gets M_2 and its associated information onto the new message list and keeps it there to the end of the present major cycle. Hence at that time the new message store holds the entry

$$M_2,\mathscr{B}(M_2),t(\mathscr{C}_1),t(\mathscr{C}_1'),t(\mathscr{C}_2).$$

At the end of the present major cycle the auctioneer makes transfers from this entry. It sends the message M_2 with the tag $t(\mathscr{C}_2)$ of its producing classifier to the old message store. And it sends the three sequences

$$+.5\mathscr{B}(M_2),t(\mathscr{C}_1);\ +.5\mathscr{B}(M_2),t(\mathscr{C}_1');\ -\mathscr{B}(M_2),t(\mathscr{C}_2)$$

to the classifier processor so that during the next major cycle it can credit the strength of the supplier classifiers $\mathscr{C}_1$ and $\mathscr{C}_1'$ and debit the strength of the producing classifier $\mathscr{C}_2$.

A basic classifier machine may be designed and programmed to recognize when a given problem, such as the object location problem, is solved. Its sensors can recognize the properties which characterize the goal state, and the classifier program can interpret these. When the goal is reached the classifier processor stops its cycling. In a bucket-brigade machine the classifier processor also rewards each classifier which has a message on the final version of the new message list.

Writing a bucket-brigade program. The Greek sophists invented a famous paradox about learning. One who learns must either learn what he knows or what he does not know. But he has no need to learn what he already knows. And he cannot what he doesn't know, for then he doesn't know what to learn!

This paradox illustrates the continuity of learning: learning begins with some knowledge and advances to more knowledge. The bucket-brigade algorithm does not create new classifiers, it only selects which classifiers are used and the circumstances under which they are used instead of other classifiers. Hence if the bucket-brigade algorithm is to produce a successful classifier program it must begin with a sufficiently rich set of classifiers.

When introducing the bucket-brigade algorithm in the last section we made some suggestions as to how one should write such a program for a particular problem. A good general approach is to formulate one's intuitive knowledge of the problem and one's hunches as to how it might be solved in a bucket-brigade classifier program. The programmer should express seemingly relevant concepts as conditions and consequents and potentially useful rules as classifiers. Since the bucket-brigade algorithm will evaluate the classifiers in a program, one need not be concerned as to whether a classifier will turn out to be useful, and can even write contradictory or contrary sets of classifiers with equanimity. The programmer may incorporate his a priori estimates of classifier strengths in the program, or he may assign equal initial strengths to the classifiers.

The use of classifier programs to express rough, intuitive, heuristic information and opinion and approximate knowledge about a problem, situation, or system is an extension of a long tradition of logical formalization and computer modeling. Logicians and analytic philosophers have used formal languages to model the rigorous aspects of language and reasoning. Computer languages are formal structures which can be employed to formulate more realistic models of language use as well as models of natural systems of all kinds.[12] Classifier languages constitute a new tool for formalization and modeling. A classifier system with the bucket-brigade learning algorithm can begin with partial information, and under the influence of information from the environment transform that partial information into precise and definite knowledge.

Evolution of a bucket-brigade program. Let us now look somewhat formally at the evolution of a bucket-brigade program from its rough initial state of partial knowledge to its state after it has learned how to solve its problem efficiently.

At the beginning of Section 7 (Learning by Repeated Competition) we considered two different basic classifier programs or sets Γ and Γ' each of which solves the object location problem of Section 6 (Transition to Learning). Let Γ^* be a set of classifiers that contains Γ and Γ' as subsets but also other classifiers, including some that are irrelevant to the object location problem. Formulate a bucket-brigade program as follows: take the set of classifiers Γ^*, assign equal strengths to these classifiers, and incorporate the bucket-brigade algorithm. Call this program Π_0^*. As it makes repeated runs it will vary, producing a sequence $\Pi_0^*, \Pi_1^*, \Pi_2^*, \ldots$.

Put this program in a robot, instruct the robot how to recognize the goal object, place the robot in the room with the object, and start it off. Because Π_0^* contains the successful subsets Γ and Γ', it will very probably solve the object location problem. However, its initial run is likely to be long and inefficient since Π_0^* contains useless classifiers as well as useful ones, and may even contain subsets of classifiers that represent conflicting and incompatible strategies.

Restart the robot each time it solves a problem, and iterate this indefinitely. As the program operates over successive trials $\Pi_0^*, \Pi_1^*, \Pi_2^*$, ... it will gradually alter its assignment of strengths to classifiers so as to improve its performance. Irrelevant classifiers will quickly lose strength and become unused, and conflicts between sub-strategies will gradually be resolved, until there is a subset of strong classifiers cooperating and competing to make a successful program.

Let Π^* be the general form of the programs $\Pi_0^*, \Pi_1^*, \Pi_2^*, \ldots$, abstracting from differences in strength assignments between successive programs. After Π^* has made enough trial runs it will learn how to solve the object-location problem efficiently, and thus will have evolved into a program with a relatively stable strength assignment which performs about the same in each run, though there will still be statistical variations. Suppose Π^* has reached this state by trial i. Let Π_s^* be the general form of the programs $\Pi_i^*, \Pi_{i+1}^*, \Pi_{i+2}^*, \ldots$, neglecting variations among them. We will call Π_s^* the "solution program" for the object-location problem.

Note that when a strength assignment is stable, the average payments made by a classifier equal the payments received by it. Hence when the solution program Π_s^* is running, the streams or flows or payments in it are invariant or steady-state. Likewise, the flows of payments between the program and its problem environment are constant.

We presented the bucket-brigade algorithm under the heading "Learning by Repeated Competition" (sec. 7). Each classifier competes on the basis of its strength to get its messages used by other classifiers, success in this enhances its strength and failure in this reduces its strength. This is an instance of the "Logic of Indirect Evalution and Selection" (sec. 3), for a classifier is evaluated through the messages it produces. This evaluation is a matter of degree, rather than all-or-none, since strength is a matter of degree. No classifier is explicitly eliminated by the process, but the strength of a classifier can become so small that the classifier is never used.

Hence the evolution of the bucket-brigade program Π^* from its initial form Π_0^* to its solution from Π_s^* is basically an evolution of its strength assignment to classifiers. The initial assignment gives the same strength to every classifier, so all classifiers start on an equal basis in the initial run. The solution program has an unequal assignment of strengths. Under it classifiers vary widely in their importance and in the roles they play in the efficient solution of the object location problem. Some classifiers have been eliminated in the sense that they play no role at all.

One example we used to present the logic of indirect evaluation and selection was a Bayesean example of learning the constitution of a box of marbles by repeated drawings from the box (sec. 3). At the beginning there is an equal assignment of prior probabilities to the alternative hypotheses. As repeated tests are made, this assignment of probabilities to hypotheses becomes unequal and focuses on one of the hypotheses. This is analogous to the change in the strength assignment to classifiers over repeated trial runs of a bucket-brigade program, though in the latter case many classifiers will emerge with high strengths. We will develop this analogy further after we have explained the genetic discovery algorithm.

Causal and economic flows in a bucket-brigade program. In each run of a classifier program there is a forward causal flow of messages through time as displayed in the series-parallel conditional produced by that run (sec. 6). Over repeated runs of a bucket-brigade program, there is also a virtual backward flow of payments to reward classifiers for their contributions to the success of these runs. These forward and backward flows occur in all bucket-brigade programs, but they are best studied in the operation of a stable classifier program like the solution program Π_s^*. By examining repeated runs of this program we can see how a classifier which contributes to the success of a run prior to the final step (so it does not share in the goal payoff for that run) later benefits indirectly from the goal payoff for that run.

Consider a series-parallel conditional generated by the solution program Π_s^* (for its structure see Fig. 8). Since Π_s^* is a solution program, we can think of this series-parallel conditional being iterated many times. To get a general picture of the backward flow of goal-payments to contributing classifiers, we will first analyze the worst case, and then consider the general case.

For the worst case we ignore the parallelism of the series-parallel conditional and concentrate on a single thread of messages and classifiers. Imagine that the following sequence occurs in a program run:

$$M_0 -- > \mathscr{C}_1 \rightarrow M_1 -- > \mathscr{C}_2 \rightarrow M_2 -- > \ldots\ldots$$
$$\mathscr{C}_{n-1} \rightarrow M_{n-1} -- > \mathscr{C}_n \rightarrow M_n,$$

where M_0 is an initial input message and M_n is an output message which brings success to the program and hence a reward (increase in strength) to $\mathscr{C}_n$. Assume further that each classifier of the sequence only supplies messages to its direct successor and never supplies messages to another classifier in this sequence.

Consider next repeated successful runs $R_1, R_2, R_3, \ldots, R_{n-1}, R_n$ of this message-classifier stream and focus on classifier $\mathscr{C}_1$'s contribution to the success of the first run R_1. Classifier $\mathscr{C}_1$ makes this contribution early in run R_1, but the goal payment does not come until the end of the run, and goes to increase the strength of a different classifier, namely, $\mathscr{C}_n$.

How, now, does classifier $\mathscr{C}_1$ benefit from this particular goal payment to $\mathscr{C}_n$? It only does so after many re-runs. The goal payment to $\mathscr{C}_n$ will affect the strength $\mathscr{C}_{n-1}$ in the next run R_2, it will affect the strength of $\mathscr{C}_{n-2}$ during run R_3, and so on down the line, until it finally affects the strength of $\mathscr{C}_1$ in R_n. Thus though classifier $\mathscr{C}_1$ ultimately benefits from its contribution of M_1 to the success of run R_1, it does not receive this benefit until $n-1$ runs later. (Classifier $\mathscr{C}_1$ may also produce several other messages during the run, and for these it will be paid earlier, of course.)

Consequently, the maximum time for a reward to influence the strength of a contributing classifier is the number of time steps from start to payoff. This is, of course, the worst case. Later classifiers in the sequence

$$M_0 \dashrightarrow \mathscr{C}_1 \rightarrow M_1 \dashrightarrow \mathscr{C}_2 \rightarrow M_2 \dashrightarrow \ldots\ldots$$
$$\mathscr{C}_{n-1} \rightarrow M_{n-1} \dashrightarrow \mathscr{C}_n \rightarrow M_n$$

receive their rewards earlier. Moreover, a typical series-parallel conditional generally has much interactive parallelism in which the same classifier functions repeatedly during the same run, so there are many payback paths from the end of a run to classifiers contributing to success early in a run. Hence the average time for payback is very much smaller than the length n, perhaps of the order "logarithm of n." In any case, repetition is essential to the success of the bucket-brigade algorithm in evaluating the contribution of each classifier to the contribution of the whole classifier program.

This example illustrates how new capital, injected into the program at the end of each successful run, flows back through the classifier program as it makes repeated runs. At each major cycle of a run successful producers pay their suppliers. In these transactions the total amount of strength or capital in the system is conserved. A goal payment adds new capital to the program. The total amount of capital in the system can be conserved by adding an external treasury from which goal payments are made, and to which classifiers make payments for input messages.

This completes our account of the flow of messages and payments in a bucket-brigade program that has solved a given problem. One can see from this account that many runs are required for goal payoffs to affect all the contributing classifiers. Let us now apply this observation to the more general and typical case, a bucket-brigade program which is learning how to solve a problem, such as the object location program Π^* as it evolves from its initial state Π_0^* to its solution form Π_s^*. The flow of messages and payments is similar in this case but more complicated.

The essence of a bucket-brigade program is a complex consisting of the classifiers with their strengths. Each of the problem solving runs $\Pi_0^*, \Pi_1^*, \Pi_2^*, \ldots$ is an experiment to test and revise this complex in the light of experience. At every major cycle a subset of classifiers cum strengths is tested and revised, and over each run the whole complex of classifiers cum strengths is tested and revised. Our analysis of the flow of messages and payments shows that many experimental runs are required for goal payments to filter back and affect all classifier operations, including those that take place early in runs.

When one also takes into account the highly contextual and holistic nature of a bucket-brigade classifier program, it is intuitively clear that very many runs are needed for the learning process. Hence fast, parallel classifier machines are needed for bucket-brigade learning.

This learning process can be accelerated by discarding poor classifiers and replacing them with more promising ones. This is what the genetic algorithm does, as we will now see.

9. LOGIC OF GENETIC DISCOVERY

A bucket-brigade program evaluates its classifiers indirectly, in terms of their consequences or products. This is one aspect of the logic of indirect evaluation and selection (sec. 3). The other aspect, that of selection, should also be applied. The classifiers of relatively low strength are of little or no use, and since they employ valuable computing resources, they should be eliminated.

Moreover, the learning power of the bucket-brigade algorithm is limited by the potentialities inherent in the initially given set of classifiers. Since this initial set is usually created by the human programmer, it is fundamentally limited by the scope of human experience. It is very desirable to transcend this limitation.

The genetic algorithm does this. It eliminates weak classifiers and replaces them by new classifiers made by genetic-type operations on strong classifiers. These new classifiers are then tested in competition with the old classifiers by the bucket-brigade algorithm.

The genetic algorithm is best introduced by means of a simple example. Suppose that at one major cycle of a bucket-brigade run the system has

Set of classifiers	$\mathscr{C}_0$	$\mathscr{C}_1$	$\mathscr{C}_2$	$\mathscr{C}_3$	$\mathscr{C}_4$	$\mathscr{C}_5$	$\mathscr{C}_6$	$\mathscr{C}_7$
Strengths	0.11	0.03	0.26	0.04	0.20	0.17	0.05	0.14

The machine might then delete the three weakest classifiers, namely $\mathscr{C}_1$, $\mathscr{C}_3$, and $\mathscr{C}_6$ and replace them by three new classifiers, derived by genetic operators from the best of the remaining classifiers, in this case, from $\mathscr{C}_2$, $\mathscr{C}_4$, and $\mathscr{C}_5$.

The genetic algorithm can use various genetic-like operators. We will illustrate it with two, mutation and crossover. Thus $\mathscr{C}_2$ might be mutated to make a new classifier $\mathscr{C}_2'$, and $\mathscr{C}_4$ crossed with $\mathscr{C}_5$ to make new classifiers $\mathscr{C}_4'$ and $\mathscr{C}_5'$. These new classifiers would be assigned average strengths of 0.125, and the strengths of the remaining classifiers reduced proportionately so that the strengths would sum to unity.

Figure 11 shows serial circuits for mutation and crossover. Both operate by the cyclic shifting of classifiers through switches that are fed from a random source. In the mutation circuit the classifier cycles through the permutation switch, which transmits it unchanged until the pulse time when it receives a random 1; then it permutes letter "#" into "0", "0" into "1", and "1" into "#". In the crossover circuit the random source produces a sequence of zeros followed by a sequence of ones. As long as the crossover switch receives 0's it returns the letters of each classifier to their original register, but when it receives 1's it crosses the classifiers over by interchanging their letters.

In embodying the genetic algorithm in a machine the designer must choose how often to apply the genetic algorithm. Biological analogies suggest letting the bucket-brigade algorithm run for several major cycles, and then applying the genetic algorithm to the existing generation of classifiers to create a new generation. But it is more efficient to have a small genetic hardware unit which processes a few classifiers each major cycle. Figure 12 shows the structure of such a machine. It consists of Figure 10 (Classifier Machine with Bucket-Brigade) augmented by a genetic processor that receives a few classifiers each major cycle from the classifier processor, applies the genetic algorithm to them, and returns the result to the classifier processor.

Genetic algorithms with mutation and crossover have been used in many classifier simulations and have proved to be very powerful. Other genetic operators, such as inversion, could also be used. One could have double classifiers, similar to diploid genetic programs, with a relation (perhaps probabilistic) of dominant and recessive holding between character strings in the processing of messages.[13] And one could use non-genetic operators to make new classifiers.

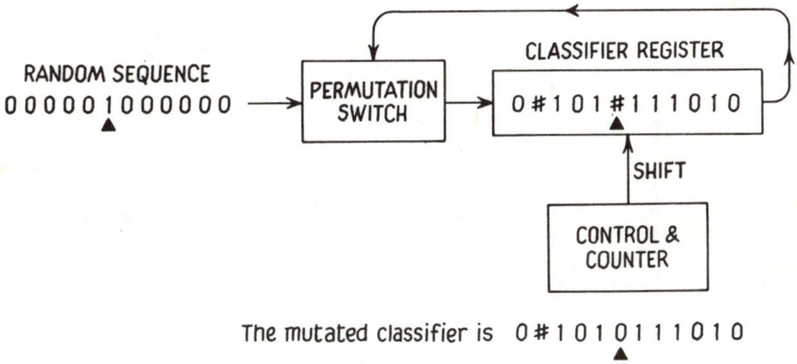

(a) Mutation circuit

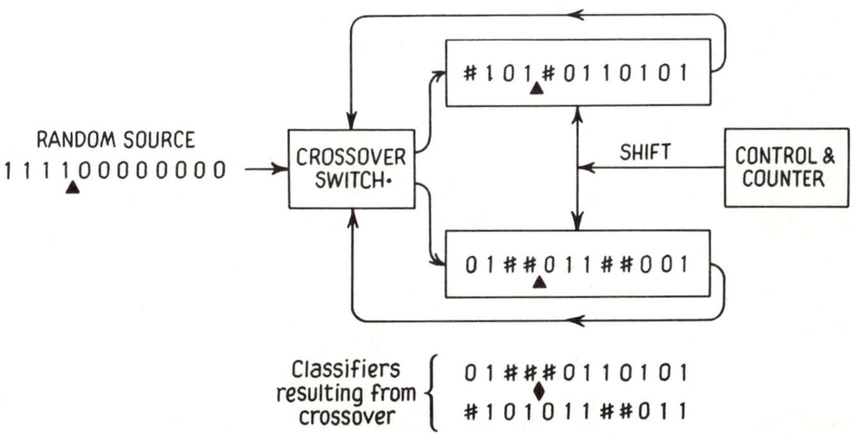

(b) Crossover circuit

Fig. 11. Genetic operator circuits for mutation and crossover.

In evaluating the use of biological analogies in classifier machines one needs to consider the contextual diferences between organic evolution and hardware design. Duplicate copies of a genetic program are relevant to biological evolution, since each copy of the same genetic program can occupy its own region of space and reproduce there. In

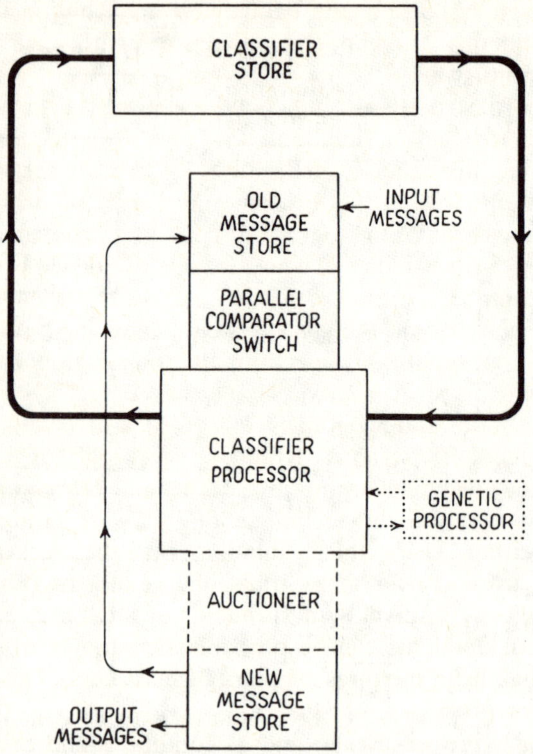

Fig. 12. Classifier machine with bucket-brigade and genetic algorithms.

contrast, extra copies of the same classifier in the machine of Figure 12 would be redundant, since all classifiers are processed in the same hardware unit. This situation could change with a different hardware organization, for example, if a classifier machine consisted of a large cellular array of chips like Figure 12, for then the cost of transmitting classifiers through the cellular space would need to be considered.

This completes our main exposition of the genetic algorithm, though we will mention some variants in a moment. We call the genetic algorithm a "logic of discovery" so let us justify the name. Logic may be divided into two branches: the logic of inquiry and discovery (what Peirce called the logic of "abduction") and the logic of argument, proof, evidence, and justification.[14] The logic of argument is now very well formalized, especially the deductive part. This logic was developed independently of computers, though computers are now used to execute portions of it.

The logic of discovery has old roots, extending back through Gottfried Leibniz to Ramon Lull. but it is only with the availability of electronic computers that formalizations of it have become possible. Classifier systems with bucket-brigade and genetic algorithms are promising formalizations of the logic of discovery. Moreover, the analyses of the present paper indicate why a really useful logic of discovery requires a computer for its execution. A discovery logic is a very complicated hierarchical feedback system, involving competition, indirect evaluation, repetition, selection, and the creation of new instructions from old ones. A discovery logic has essential connections to learning, evolution, and economics, all of which deal with complicated hierarchical feedback systems. I think a good logic of discovery will also be a good logic of learning, evolution, and economics.

Let us now analyze the technique of discovery employed by the genetic algorithm. Using the evaluations that the bucket-brigade algorithm has learned over many trials, the genetic algorithm replaces poor and useless classifiers by descendents of the most valuable classifiers. These descendents can be probabilistic variants of single classifiers or probabilistic mixtures of pairs of classifiers. This is a guided trial and error process.

A classifier expresses a rule, and its conditions and consequent are terms expressing concepts that represent properties. Hence the genetic algorithm creates new rules and new concepts. Moreover, a rule is an applied conditional statement, the rule and the statement having the same informational content, so that the genetic algorithm also discovers new statements. Thus every classifier is a tentative hypothesis, and the genetic algorithm constantly generates new hypotheses for the bucket-brigade algorithm to test.

Hierarchical structuring is important in complex systems, so the hierarchical difference between mutation and crossover is important. Mutation changes single letters, and hence creates new sequences of letters. Crossover rearranges sequences of letters, and if these sequences express concepts then crossover rearranges concepts. An example is shown in Figure 13. Operating on two rules which make the wrong correlations between the color of a stop light and the actions of stop vs. go, crossover might produce the correct rules. More generally,

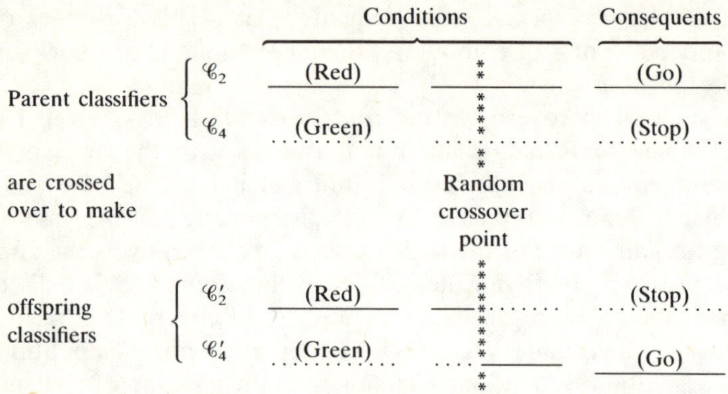

Fig. 13. Crossover in the genetic algorithm.

crossover may associate two mutually supportive properties that were not associated in the parent classifiers. The difference between producing new sequences of letters (mutation) and producing new sequences of sequences of letters (crossover) is a difference in hierarchical level.

We saw in the last section that if a bucket-brigade program is to evolve into a program that solves a given problem, then the initial program must begin with a sufficiently rich set of classifiers. This limitation is removed by the genetic algorithm, because that algorithm produces new classifiers. A classifier program with both algorithms can begin with a random set of classifiers and evolve into a successful program. If the initial set of classifiers fails to respond to any message, new classifiers which do respond can be generated by crossing old classifiers with input messages.

We have now presented the essentials of classifier systems and their learning and discovery algorithms. There are many details of implementation which we have not attempted to cover, and there are many important variants and ramifications of classifier systems. We will mention a few but refer the reader to the bibliography for further information.

We explained earlier how to write a bucket-brigade program (sec. 8). The same procedure applies when the genetic algorithm is added (sec. 9), though with it one can write the initial classifiers at random. To simulate a classifier program on a computer one needs to express the environment as a program. The environment could be written as a basic classifier program. Then the whole system would be expressed as a classifier program.

Consider next the role of negation in classifier systems. The only use of negation inside the system is on the second condition of a classifier, with the effect that a message satisfying the first condition only leads to a new message when *no* message satisfies the second condition. For example, with the classifier "If green and not-red, then move forward," a green stimulus produces motion only if no red stimulus is simultaneously present. Hence contradictions cannot appear inside the system. But classifiers can produce incompatible output messages, such as "move forward" and "stand still." Hence some means are needed for presenting contradictions at the output. This can be accomplished by having classifiers which produce output messages compete in a bucket-brigade auction to get their messages sent to the output interface.

Classifiers occupy valuable memory space and use processor capacity, so they should be taxed for these. We suggested in connection with the bucket-brigade algorithm that there be an external treasury which would receive payments for input messages and make payoffs for goal attainment (sec. 8). Taxes could be deposited in this treasury so that the total amount of capital in the system would be constant.

We described the external payoffs made by the bucket-brigade algorithm as if they were all positive (rewards). This was for the sake of simplicity. There could also be negative payoffs, punishments or penalties levied each time a classifier program fails. We explained how the bucket-brigade algorithm distributes rewards to the whole program to its parts, the classifiers, and the same explanation covers the distribution of penalties. When a classifier program recognized that it had failed, it would charge those classifiers which were active at the time. Over successive runs these losses would filter back to the classifiers which contributed to this failure early in the run.

Classifiers with bucket-brigade learning and genetic discovery are a novel type of intelligent system that combines features from both artificial computing systems and natural intelligent systems. Ideas for extending classifier systems can also be taken from both the artificial and the natural realms. We give some examples.

A classifier is a special kind of rule, and other kinds of rules could also be used. The standard arithmetic and logical transformations executed by computers could be expressed as rules operating on messages and added to classifier systems. All these rules are like finite automata transition functions in being determinate. As we said in Section 6, one could also employ probabilistic classifiers, which resemble the transition function of probabilistic automata.

In the bucket-brigade algorithm the bid a classifier makes for its message takes account of the classifier's strength and the specificity of its conditions. There may be other factors which should be taken account of. One is the strength of *association* of a classifier with the classifiers that supply it with messages. Consider two classifiers $\mathcal{C}_1$ and $\mathcal{C}_2$ competing with their messages M_1 and M_2 in a situation where they would make the same bid under the formula

$$\mathcal{B}(M) = \text{constant} \times \text{strength}\ (\mathcal{C}) \times \text{specificity}\ (\mathcal{C}).$$

But suppose further that in this situation $\mathcal{C}_1$ is associated with or supported by more classifiers than $\mathcal{C}_2$ in the following sense. Many messages satisfy the conditions of $\mathcal{C}_1$ and the sum of the bids made to get them into this major cycle is substantial, whereas only two messages satisfy the conditions of $\mathcal{C}_2$ and the sum of bids made to get them into the major cycle is relatively small. Thus many classifiers have supplied $\mathcal{C}_1$ with messages and are therefore relevant to it, and their "support" (the sum of their bids) is large. In contrast, the support of $\mathcal{C}_2$ is small. These considerations show that *support* ($\mathcal{C}$) should also be a factor in the bid formula, and simulations have borne this out.

Since classifiers are instructions and the genetic algorithm modifies them, the genetic algorithm is a meta-device or metalanguage algorithm working on the object language. There are lots of alternatives and generalizations possible here. There could be metalinguistic operators, other than genetic operators, for making new classifiers. Also, the system could select sets of interrelated classifiers and make them the objects of the genetic operators, for any of various kinds of interrelations detectable by a classifier program. Compare the genome as a string of genes and Peirce's notion of a "bundle" of interacting habits. Or, as in nature, the system could establish competition between groups of interbreeding classifiers. Various of these procedures could lead to the development of hierarchies within the system.

Our last point is on probability, one of the subjects of this conference. We mentioned (sec. 7) that probability is used in the bucket-brigade auction so that messages with low bids have some chance of being carried over to the next major cycle. Such messages might be useful and therefore it is a poor strategy to exclude them deterministically. For the same reasons, probability should be used in the genetic algorithm (sec. 9) to select the classifiers that are eliminated and to select the classifiers that are mutated and crossed over. Mutation and crossover are already probabilistic operators. Thus a classifier program with bucket-brigade learning and genetic discovery is a stochastic process.

We go now to discuss the other subject of this conference, causality.

10. EVOLUTION, INDUCTION, AND CAUSALITY

We have been using the object location problem of Section 6 as a paradigm for the study of learning and discovery. It is a geographical problem of mapping out the positions of objects in space, and hence involves learning a group of specific facts. This is a restricted type of problem. The logic of evolution also applies to general statements, so classifiers can be used to discover and learn about causal connections, both determinate causal connections and probabilistic causal connections (empirical probabilities).[15]

Whle causality sometimes functions as a constraint on the simultaneous occurrence of properties, usually it involves the temporal sequence of cause followed by effect. To facilitate the discovery and verification of causal sequences, one could augment classifier systems so they would look for and directly recognize temporal successions. Classifiers themselves constitute causal rules connecting messages of one major cycle to messages of the next, but cause and effect in nature may be separated by several major cycles, in which case long-term coupling between classifiers would be needed. This can be achieved by keeping a record of classifiers that are active at successive major cycles and making new classifiers from them.

One philosophical approach to the problem of causality is to analyze the nature of inductive and probabilistic arguments from observed data to assertions of causal connections. And one answer within this approach is a Bayesean model of confirmation. The idea of using Bayes' theorem of inverse probabilities as the basis for a logic of induction is very old. Bayeseanism has continued to have appeal because a Bayesean model captures several common features of actual inductive inference: the presence, at least in the background, of alternative hypotheses; testing hypotheses by their consequences and adopting a well-confirmed hypothesis; the feature that confirmation is a matter of degree.

In *Chance, Cause, Reason* I presented the Bayesean model of induction as a method of finding an unknown probability, and grounded it on a theorem about the length of experiments, which I called the "confirmation theorem." Expressed informally this theorem states that an investigator who applies Bayes' method of finding an unknown probability to a mixture of independent processes will eventually know the truth. This theorem combines Bayes' theorem with the law of large numbers, and implies as a corollary that two investigators who disagree only on their prior probability assignments will after repeated experimentation come to agree on the true hypothesis. This Bayesean model is closely related to Rudolf Carnap's confirmation functions and Bruno de Finetti's concept of exchangeability. The Bayesean model can also be extended to deal with causally necessary generalizations.[16]

In the preceding section of the present paper we distinguish between the logic of argument and the logic of inquiry and discovery. The Bayesean model is only intended as a model of inductive argument, not as a model of inquiry and discovery. I will argue now that the logic of bucket-brigade learning (secs. 7 and 8) and the logic of genetic discovery (sec. 9) constitute natural and coherent extensions of a Bayesean model of inductive argument (cf. sec. 3).

As a first step we compare the operation of a Bayesean confirmation process with the operation of the bucket-brigade algorithm.

In the Bayesean model there is a starting set of hypotheses which remains invariant during the process. These hypotheses compete in the strong sense that only one will be selected as well-confirmed. The hypotheses are repeatedly tested in a sequence or run of experiments that terminates when one hypothesis is sufficiently confirmed. At each step or major cycle of the run there is an assignment of weights or probabilities to the hypotheses. The initial values are called "prior probabilities" and subsequent values are called "posterior probabilities." The sum of these weights is constant, and is usually taken to be one at each step. After a sufficient number of experiments or trials this weight assignment will stabilize within various limits, with a peak at one hypothesis. When this peak is sufficiently close to one the run will have reached its goal of finding a well-confirmed hypothesis, and will stop.

The operation of a classifier system with the bucket-brigade learning algorithm is similar to this Bayesean process, but is much more complex. There is a starting set of classifiers, which remains invariant during the learning process. These classifiers both compete and cooperate with one another, and at the end of the learning process the problem-solving role of each, if any, will be settled. The given set of classifiers is tested repeatedly at two temporal levels: there is a sequence of runs, each a sequence of major cycles.

At each step or major cycle there is an assignment of weights or strengths to the classifiers. Initially this is a given assignment. During a major cycle the strength assignment changes as a result of the bucket-brigade auction. Furthermore, payoffs are made at the end of each run, a reward for success, a charge for failure. These payments come from or go to a treasury, which also receives payments for input messages. Taken over the whole system, the sum of strengths is constant, and can be normalized to unity, as in the case with probabilities. After a sufficient number of successful runs the classifier program will have learned to solve the problem efficiently, and the strength assignment to classifiers will stabilize. This steady-state assignment determines the degree to which each classifier participates in a problem-solving run, if at all. It is thus a parallel selection function, producing a multi-peak strength assignment.

When the genetic discovery algorithm is added to a bucket-brigade classifier system the process becomes more complex in that the set of classifiers change, weak classifiers being replaced by genetic combinations of strong classifiers.

There is thus a natural development from Bayes' method of finding an unknown probability to bucket-brigade learning and on to genetic discovery. All of these procedures are learning and evolutionary processes in the most general sense. The Bayesean procedure learns the value of an unknown probability, and the classifier systems learn to solve more general and more complex problems. Each procedure can be used to solve artificial problems and to learn about nature and the environment.

Since classifier learning and discovery procedures constitute more advanced forms of probabilistic induction than Bayes' method of finding an unknown probability, they should be useful in resolving some traditional problems of induction. One of these problems concerns the relative roles of enumerative induction and eliminative induction, so let me address that issue.

Enumerative induction is illustrated by Bayes' method of finding an unknown probability. It starts with given properties and tests their association by repeated instances. Each instance counts in the sense of altering the posterior probabilities, though if the probability of one hypothesis is close to unity after a long run of experiments, further instances are unlikely to make much difference.

Eliminative induction begins with a set of properties believed to be relevant to the problem at hand, such as the problem of finding the cause of a phenomenon. To ascertain which properties are causally related the investigator varies some of the properties while keeping the others fixed. He hopes in this way to find the causal law which will explain the phenomenon. Francis Bacon's tables of instances and Mill's methods of experimental inquiry are both examples of eliminative induction. The controlled experiment is a quantitative version of eliminative induction.

One can ask of these two forms of induction, which describes scientific practice more accurately and which is more basic?[17] Let us examine these issues from the point of view of classifier learning and discovery.

Our discussion of the evolution of a bucket-brigade program from its initial form to its solution form showed that repeated experimentation is basic to classifier learning (sec. 8). Each time a classifier generates a message, that classifier with its strength is tested in the context of other classifiers with their strengths. Many runs are required before a payoff that is made at the end of a run works its way back and influences classifier actions that took place early in earlier runs. Since a bucket-brigade classifier system is highly contextual and holistic, many major cycles are required for the initial form of a program to evolve into the solution form. When the genetic algorithm is added to the bucket-brigade algorithm the classifier system learns faster, but if it begins with a poor initial set of classifiers it will take a long time to learn. For these reasons, fast parallel classifier machines are needed for learning and discovery.

This proposed machine parallelism should be distinguished from the intrinsic sampling parallelism of the classifier learning and discovery algorithms. This sampling operates on two levels. First, a condition word with "don't cares" is general, being satisfied by a class of messages. Each time the bucket-brigade algorithm tests a classifier it is sampling all the messages that satisfy the conditions of that classifier. Second, an individual classifier is a sample of the set of classifiers derivable from it by the genetic operators. Hence, in creating new classifiers by the genetic algorithm and testing them by the bucket-brigade algorithm, a classifier program is sampling sets of classifiers.

All these considerations show that repetition and the enumeration of cases is basic to classifier learning and discovery, and hence to the logic of evolution generally. Nevertheless, in its more advanced stages commonsense learning and scientific inquiry often settle questions in the almost all-or-none way of eliminative induction, and experiments are only repeated enough so the investigators can be sure that the controls were operating as planned and that the observations are repeatable. It is of interest to learn how this type of induction might develop in classifier systems.

When an investigator applies eliminative induction to some general subject-matter domain, he already has an established body of probable knowledge concerning the laws he is likely to find and the properties involved in them. He has probabilistic information as to what properties are likely to be interconnected by the causal laws operating in the domain, and also as to which combinations are worthy of test. He also has available established laws and procedures for use in planning his tests. I think this knowledge has evolved in a hierarchical manner from lower levels of knowledge which depend more heavily on repetition and enumeration than the higher levels.

Let us now speculate on how this kind of hierarchy might evolve in classifier systems. Note that the genetic algorithm modifies the conditions and the consequents of classifiers, as well as the connections between these. Looked at linguistically or logically, conditions and consequents are terms referring to properties, either properties of the environment or properties used in dealing with the environment. Classifiers are statements expressing generalizations and rules. Hence a classifier system with the bucket-brigade and genetic algorithms is searching for new concepts that are likely to be useful in dealing with the environment and solving problems, as well as for combinations of these concepts into statements that express true generalizations and useful rules.

In order for a hierarchical knowledge structure to develop in a knowledge system, the concepts, generalizations, and rules of one level need to become building blocks for more complex concepts, generalizations, and rules at the next level. An analog from biological evolution illustrates this point. Cells are low-level building blocks in evolution. But once stable clusters of cells such as organs, skin, and bones have emerged, the evolutionary process uses them as building blocks for new organisms, without having to design these organisms completely from lower-level cells.

How might these higher level building blocks emerge in a classifier program? One mechanism was suggested near the end of the previous section and early in this section: the genetic algorithm could be applied to sets of classifiers interrelated in some way ascertainable by the classifier program. Another mechanism is inherent in the role of specificity in calculating bids. There is an old rule in informal logic that "every rule has exceptions." Classifiers can discover and handle exceptions, in the following manner.

For simplicity of exposition let us consider one-condition classifiers. Let $\mathscr{C}_1$ be a classifier with condition ####0000 which works well for all messages which satisfy it except for messages 0111000 and 11110000. For these two messages it wins bids and gets its products carried over to the next major cycle but these products are not used, so $\mathscr{C}_1$ loses money on them. Let $\mathscr{C}_2$ be a more specific classifier with condition #1110000 which works well for these two messages 01110000 and 11110000. Assume that $\mathscr{C}_1$ and $\mathscr{C}_2$ are of comparable strength. Since $\mathscr{C}_2$ is more specific than $\mathscr{C}_1$ it will outbid $\mathscr{C}_1$ in the exceptional cases where $\mathscr{C}_1$ fails, thereby helping $\mathscr{C}_1$ as well as itself.

What happens inside a classifier system as it operates depends on the environment as well as the classifier system. Let us suggest an environment under which a suitable classifier system might develop a hierarchy. Consider once again the object location problem we have used to illustrate classifier learning and discovery (sec. 6). Generalize it by defining several other problems that are similar to it in important respects but also different from it in important respects. Then let the classifier program which solved the object location (the program Π_s^* of sec. 8) work on these problems. I think it likely that this program would develop higher-level concepts, generalizations, and rules that would transfer learning from some of these problems to others.

We have now compared and contrasted three methods which represent successive stages of the logic of evolution: Bayes' method of finding an unknown probability, classifiers with bucket-brigade learning, and classifiers with both bucket-brigade learning and genetic discovery. In conclusion we will discuss how each of these methods converges on some truths about the environment.

In Bayesean induction this convergence is quite simple. The method presupposes that the environment is characterized by one of a given distribution of empirical probabilities. Bayes' method of finding an unknown probability consists in performing successive experiments, observing the outcomes, and calculating the posterior probabilities of the alternative hypotheses on the basis of the accumulated evidence. By the confirmation theorem the method will ultimately converge on the true hypothesis, that is, the correct value of empirical probability.

Moreover, assuming a prior probability assignment to hypotheses, one can estimate in advance the number of experiments needed to approximate the truth to some specified accuracy. More technically, given a threshold probability for hypothesis acceptance, for each sequence of N experiments one can calculate the likelihood of the experimenter accepting some hypothesis at the end of the sequence, and one can calculate the probability that the accepted hypothesis is true. Thus, starting from a prior probability distribution and some criteria for hypothesis acceptance and degree of certainty, one can calculate the number of experiments needed to learn the approximate truth about the environment.

The corresponding issues for classifier learning and discovery are, of course, much more complex and problematic. The Bayesean result assumes that the environment is fixed. A good learning and discovery system should be able to follow or "track" some changing environments, but the problem of learning about a steady-state environment is already very difficult. Consider a classifier program which is intended to learn about some characteristic or law of the environment. After a sufficient number of runs of various lengths it may evolve into a program which has solved the problem and learned about the environment. There are several interesting questions which should be investigated.

Under what conditions will the program solve the problem and learn about the environment? How long will it take to do so, how many runs and how long will the runs be? When the program reaches some kind of stability what will it be like – a single point as in the Bayesean case, a periodic oscillation, or some very complex behavior like the "strange attractors" found in the study of dynamic non-linear systems? Can one use results about the complexity of programs and the lengths of experimental runs to measure the complexity of problems and environments? The answers to these questions are obviously different for classifiers with bucket-brigade learning only and for classifiers with bucket-brigade learning and genetic discovery.

The general theory of truth implicit here is a pragmatic correspondence theory. An important factor in the evolutionary survival of an adaptive system is the "fit" or "impedance match" of the system's abilities, habits, and beliefs to the corresponding properties and laws of the environment and other adaptive systems. As lower organisms evolved into rational man, lower-order responses evolved into intentional thought and rational action. True statements are useful because the properties and laws expressed by them correspond to the properties and laws of nature.

11. THE REDUCTION OF HOLISTIC-COHERENT SYSTEMS TO HIERARCHICAL-FEEDBACK SYSTEMS

The main thesis of this paper is that holistic-coherent systems are in principle reducible to hierarchical-feedback systems. This thesis follows from my general philosophy, the philosophy of logical mechanism, which holds that all systems are so reducible. To support this philosophy I have argued elsewhere that man is a finite automaton and, more generally, that natural processes are simulable by computer. Furthermore, I have given computer accounts of human intentionality and functional consciousness, and have suggested a von Neumann type robotic model of evolution.[18]

In the present paper I have developed a logic of evolution based on a Bayesean model of induction and Holland's bucket-brigade learning and genetic discovery algorithms for classifier systems. This logic of evolution is the main premise of a further argument for the reduction of holistic-coherent to hierarchical-feedback systems.

Natural evolution is clearly a fundamental type of hierarchical-feedback system. It is complex, dynamic, and non-linear. It has many fast-operating and interacting feedback cycles which make its behavior sensitive to small effects. It has both analog and discrete features and involves information flow as well as materials flow. It develops hierarchies, which are important for its organization and its rate of development.

My further argument for the reducibility of hierarchical-feedback systems to hierarchical-feedback systems is as follows. The logic of evolution that we have sketched here is realizable in automata (computers and robots). Automata are hierarchical-feedback systems. Therefore, insofar as our logic captures the essence of natural evolution, it shows that this very basic holistic-coherent system is reducible to a hierarchical-feedback system.

Moreover, starting from a physical and chemical base, natural evolution has produced many complex holistic-coherent systems, such as organisms, species, societies, and cultures. Since these systems have been produced by a hierarchical-feedback mechanism, they must themselves be reducible to hierarchical-feedback systems. This is a strong analogical argument for the conclusion that all holistic-coherent systems are so reducible.

The logic of evolution also strengthens the case for the reducibility of holistic-coherent systems by explaining how an important and broad class of these systems function. We have mentioned some examples in the process of elucidating the logic of evolution in general and classifier systems in particular. We will further illustrate this point by applying the logic of evolution to two philosophical doctrines about holistic-coherent epistemological systems: the Hegelian thesis that knowledge develops by an evolutionary process of iterated opposition and synthesis; and the coherence theory of truth.

G. W. F. Hegel took the development of human culture as his model for all systems and formulated his dialectical method as the fundamental logical law governing the operation of systems. Culture is a holistic-coherent system, and Hegel's philosophy was the source of many holistic-coherent philosophical epistemologies, including absolute idealism, Marxism, and John Dewey's pragmatism.

Each application of the dialectical method has three stages: Thesis, Antithesis, and Synthesis. A Thesis produces a contradicting Antithesis, and the incompatibility of Thesis and Antithesis leads to a Synthesis that contains the best of both Thesis and Antithesis. This process is iterated, often producing new levels of a hierarchy.

It is important to understand the nature of contradictions in Hegelian dialectic before comparing this dialectic with our logic of evolution. A dialectical contradiction is not an explicit contradiction of the form $p \& \sim p$. Rather, a dialectical contradiction is an intellectual opposition of seemingly incompatible theses that are capable of modification and reconciliation in a higher synthesis. We give an example from the history of science. It seemed for centuries that the wave and corpuscular theories of light were logically incompatible, so both could not be true. This opposition was a positive stimulus to research, and the apparent contradiction was removed when both theories were reduced to quantum theory. An explicit logical contradiction can lead to an unlimited number of other explicit contradictions, but the conjunction of the two theories of light did not do so.

There are some similarities between classifier systems and Hegelian dialectic which bear on the question of whether Hegelian coherence can be explained by the logic of evolution and is therefore reducible. Of course, Hegelian dialectic and classifier systems are on very different levels of information processing. Dialectic is a vaguely formulated method for dealing with broad and high-level theses, whereas classifiers are low-level systems with simple but precise rules operating on computers. Moreover, dialectic logic is determinate and serial, whereas classifier systems are probabilistic and parallel.

But conflict and cooperation are fundamental in both systems. In dialectical logic there is the conflict of Thesis with Antithesis; in a classifier system with bucket-brigade learning, classifiers both compete and cooperate. Dialectical logic combines parts of a Thesis and its Antithesis that work well together to make a Synthesis that is superior to both Thesis and Antithesis. The crossover operator of the genetic algorithm combines parts of two classifiers that *may* work well together, and if these parts to work well together the new classifiers will be stronger than either parent classifier.

I think that Hegel saw correctly that rational conflict and cooperation play important roles in the development of knowledge, though his formulation of the process is very rough and informal. I think also that a more formal and precise account of the conflict and cooperation of ideas and theories could be given by means of classifier programs. Insofar as this is plausible, it constitutes an argument for the reduction of Hegelian dialectical systems, a prominent class of holistic-coherent systems, to hierarchical-feedback systems.

This completes our brief discussion of the Hegelian thesis that knowledge develops by an evolutionary process of iterated opposition and synthesis. We turn now to the coherence theory of truth, which is generally contrasted with the correspondence theory.

The idea of coherence is a difficult one, so we must first discuss what it is. I find three distinguishable theses that coherentists seem to advance. First, that knowledge systems are highly complex, well-integrated, and hierarchical, and are not mere collections of elementary statements. Second, that in a well-established knowledge system the higher-level theories do not correspond to objective facts in the simple and direct way that true observation statements and low-level generalizations correspond to facts. Third, that only a complete scientific knowledge system can be verified or justified, and not single scientific theories or statements. This last thesis was formulated by Pierre Duhem, but it is at least implicit in most coherence theories.

The first thesis is borne out by the analyses of this paper. The second thesis seems true, but is compatible with a sophisticated correspondence theory of truth. Both theses fit well with the philosophy of logical mechanism. Hence we need only discuss the third thesis.

Since this is a whole-part issue about scientific knowledge systems, let us look at the composition of these systems. They typically have a hierarchical structure. Statements and rules are composed of terms. Related groups of statements constitute theories, and related groups of rules constitute methods and procedures. A subject matter is a system of theories and procedures. A statement is tested in the context of other statements. The test of a theory presupposes the truth of auxiliary theories and the validity of background procedures.

We give an historical example. When Jean Foucault showed that the velocity of light in water is less than in air, a result that favored the wave theory of light over the corpuscular theory, he assumed the theory of gravitation, the laws of the reflection and refraction of light, and the validity of many experimental procedures.

The coherence thesis that justification is of a whole knowledge system rather than of its parts applies to each level of the system. Thus, single statements cannot be justified alone, but only systems of statements (or theories); and isolated theories cannot be justified but only holistic groups of theories; etc. Expressed in terms of the concepts of the present paper, the coherence theory holds that the empirical justification of an epistemological whole cannot be allocated to, or reasonably distributed among, the component parts of the whole.

The opposing thesis is that the empirical justification of a theory or system of theories can be distributed among the component statements or theories. How might this be done? It clearly cannot be done in terms of deductive interrelations between observation reports, statements, and theories, because deductive relations are all-or-none. But it can be done in terms of inductive interrelations. The inductive justification of a knowledge system can be distributed among its constituents in terms of probability connections and the inductive strengths of the components.[19]

In the last section we showed some similarities between a Bayesean model of inductive verification and bucket-brigade learning. This connection suggests that the bucket-brigade method of distributing payoffs is relevant to the coherence issue. The suggestion is strengthened by the fact that a classifier system with bucket-brigade learning is itself a kind of empirical knowledge system, since the classifier system is learning about the problem environment in which it is placed.

Let us compare the successful operation of a classifier system with bucket-brigade to the process of testing and improving a scientific knowledge system. A single classifier does not determine whether the messages satisfying its conditions produce messages that are carried over to the next major cycle. Rather, this is determined probabilistically by the interaction of the classifier cum strength with the rest of the classifier-strength pairs of the system. Still, when a bucket-brigade program receives a reward for success it gradually allocates this to the contributing classifiers.

By analogy, an investigator does not deduce an observable prediction from a single theory, but rather derives it (usually probabilistically) from the theory together with background theories and statements. When the prediction is tested the investigator estimates the probabilistic bearing of it on both the main theory under test and the contextual assumptions, and on this basis modifies the system so as to better fit the data and the world. His probability estimates are vague and subjective, but this is to be expected, because a scientific knowledge system cannot be applied many times in order to allocate probabilities to its different constituents, in the way a bucket-brigade program can be run many times to allocate strengths to its classifiers.

I think this analogy between the evolution of a scientific knowledge system and the evolution of a classifier system helps show that single empirical statements and theories can be verified or justified, and hence counts against the coherence theory that only complete knowledge systems can be verified or justified.

Our analogy of a system of scientific theories to a classifier system illustrates the broad perspective with which the philosophy of logical mechanism views systems and system design. A system is an hierarchical organization or relational structure of building blocks or constitutive parts, and these parts may be abstract or symbolic as well as concrete. Many possible systems can be constructed from a given repetoire of parts. Associated with each such construct is a cost, which is a function of the costs of its parts and of their assembly or organization. Systems operate in environments, and when a criterion of achievement is defined, the performance of a system in an environment can be rated. On this basis systems can be compared with respect to their utilities. (An environment is also a system, of course, but we are discussing the case where the environment is changing less rapidly than the system evolving in it.)

We will illustrate the economy of possible systems further with a comparison of computer architectures to scientific knowledge systems. Given a mix of problems to be solved and the costs associated with computer parts and their assembly into complete computers, computer architectures can be evaluated in terms of their performances and costs. Analogous questons can be asked about scientific knowledge systems. How well does the system perform in an environment; that is, how well does the system fit the experimental data and explain the phenomena? How simple is the system, and easy to work with? The performance of a scientific knowledge system depends both on the performance of its constituent parts and its organization.

This completes our case that holistic-coherent systems are reducible to hierarchical-feedback systems. We will give a brief summary.

Our main tool has been a proposed logic of evolution, which is designed to formulate the rationale of evolutionary processes generally. The logic of evolution has three parts: logic of indirect evaluation and selection, learning by repeated competition, and logic of genetic discovery. These have been developed at length in terms of a model of induction based on Bayes' theorem of inverse probabilities, Holland's classifier systems with bucket-brigade learning, and classifier systems with both bucket-brigade learning and genetic discovery. These classifier systems have a design procedure built into them, as biological evolution clearly does.

Classifier systems with learning and discovery are computer systems, and hence logical-mechanical systems. But it is important for the thesis of this paper that they are also holistic-coherent systems. Over successive major cycles, a classifier interacts via the auction system with the other classifiers, and its strength depends on its performance in competition with other classifiers, each with its strength. And over successive problem runs, the strength of a classifier depends on the payoffs received by the whole system as distributed by bucket-brigade feedback. These interactions all take place at the object-language level. In addition, there are interactions at the meta-level. At the meta-language level the construction of new classifiers depends on the relative strengths of all the classifiers. Because they are holistic-coherent systems, classifiers are useful models of such systems and the basis of a strong argument for the reduction of holistic-coherent systems to hierarchical-feedback systems.

Since this is a conference on causality, I will conclude with an analysis of the causality of evolution. The essential point is that the logic of evolution works *indirectly*, even though it operates in a logical-mechanical system of *direct* or non-backwards causality.

The logic of evolution applies to a complex consisting of an evolving system interacting with an environment. The evolving system contains producers which repeatedly generate products. These products are evaluated competitively by the environment and by the evolving system itself. The evolving system uses these evaluations to select the degree to which each producer contributes to the operation of the system and to generate new producers. Insofar as it is successful the evolving system increases the rewards it receives, that is, its fitness to the environment.

Each of these constituent evolutionary operations is accomplished by means of direct causality. Still, the producing entities of an evolutionary system are evaluated, selected, and modified according to the nature of their products or consequences. Consequently, the causality of the evolutionary process as a whole is indirect. By analogy with Bayes' theorem of inverse probabilities, we propose to call this holistic kind of causality *inverse causality*. Inverse causality underlies induction, learning, discovery, teleology, design, adaptation, and evolutionary processes generally. We have shown in this paper how to reduce inverse causality to direct causality. Thus, inverse causality is the foundation for the reduction of holistic-coherent systems to hierarchical-feedback systems.

Department of Philosophy
University of Michigan

NOTES

* The work reported herein, and the writing of this paper, were supported by the National Science Foundation grant DCR-8305830 and predecessor grants awarded to the author and John Holland. See the Holland papers in the bibliography. I wish to thank John for the benefit of many discussions of his classifier systems.

This paper was presented at the conference on Probability and Causation, University of California at Irvine, July 16, 1985.

[1] Thomas Kuhn, *The Structure of Scientific Revolutions*, is a current example. See Israel Sheffler, *Subjectivity in Science*, for a good analysis of several holistic and relativistic treatments of science. The late Richard McKeon held that each well-developed philosophical method-and-system was internally coherent and relatively impenetrable by opposing systems.

[2] This will be the title of a forthcoming book of essays on my philosophy, being edited by Merilee Salmon and to be published by Reidel.

[3] 'Logic, Computers, and Men'; *Chance*, *Cause*, *Reason*, sec. 10.4.1; 'Computers, Control, and Intentionality' and 'An Architectural Theory of Functional Consciousness'.

[4] 'Computers, Control, and Intentionality', pp. 44, 53; 'An Architectural Theory of Functional Consciousness', secs. 5–7.

[5] *Chance*, *Cause*, *Reason*, sec 9.4.1. 'Computer Science and Philosophy', sec. 3. For a more formal treatment see my 'Models of Deterministic Systems'.

[6] See my 'Computer Science and Philosophy', sec. 3. See also my discussion of Bernard Ziegler's essay 'Cellular Space Models: New Foundations for Simulation and Science', to be published in *The Philosophy of Logical Mechanism*, edited by Merilee Salmon.

[7] See my *Chance*, *Cause*, *Reason*, sec. 2.5.

[8] See the items in the reference list by Holland, Holland and Reitman, Holland and Burks, Bethke, Booker, DeJong, Forrest, Gillies, Goldberg, Grefenstette, Skolnick, and Smith. The theoretical background for classifier systems is developed in Holland's *Adaptation in Natural and Artificial Systems*.

[9] Classifiers have been applied to pattern recognition problems by Skolnick ('Automated Comparison of Image Similarities and Differences') and Gillies ('Machine Learning Procedures for Generating Image Domain Features').

[10] For descriptions of the first three, see Goldberg, 'Computer-Aided Gas Pipeline Operation Using Genetic Algorithms and Rule Learning'; Smith, 'A Learning System Based on Genetic Algorithms; and Gillies, 'Machine Learning Procedures for Generating Image Domain Features'.

[11] See Savage, *The Foundations of Statistics*, secs. 1–3 of ch. 2 and secs. 1–3 of ch. 5. See also my 'calculus of choice', ch. 5 of *Chance*, *Cause*, *Reason*, especially the main theorem (p. 288).

[12] See my 'Logic, Biology, and Automata – Some Historical Reflections', pp. 308–311 and 'Computer Science and Philosophy', pp. 408–410.

[13] See Holland, *Adaptation in Natural and Artificial Systems*, pp. 112–116, 119.

[14] This division is orthogonal to that between deductive and inductive logic. For a discussion of kinds of logic see my *Chance*, *Cause*, *Reason*, sec. 1.4.

[15] See my *Chance*, *Cause*, *Reason* for a discussion of the nature of causal connections (ch. 7) and the nature of empirical probabilities (ch. 8) as contrasted to inductive probabilities (ch. 5).

[16] For a full explanation see my *Chance*, *Cause*, *Reason*, secs. 2.5 and 3.2. In 'Induction and Causality in a Cellular Space', Sohichi Uchii has shown how my general procedure for confirming causally necessary statements can be applied to cellular spaces.

[17] See Cohen's 'What Has Inductive Logic to Do with Causality?' my 'Enumerative Induction versus Eliminative Induction', and the comments following these two papers.

[18] See the references in sec. 2, and my 'Von Neumann's Self-Reproducing Automata'.

[19] This is the thesis of my paper 'Justification in Science'.

BIBLIOGRAPHY

Bethke, A. D. (1980) 'Genetic algorithms as function optimizers.' Doctoral thesis, Computer and Communication Sciences, University of Michigan.

Booker, Lashon (1982) 'Intelligent behavior as an adaptation to the task environment.' Doctoral thesis, Computer and Communication Sciences, University of Michigan.

Burks, Arthur (1953) 'Justification in science', *Academic Freedom, Logic, and Religion*, pp. 109–125. Morton White (ed.), Philadelphia: University of Pennsylvania Press.

Burks, Arthur (1970) 'Von Neumann's self-reproducing automata', pp. 3–64 of *Essays on Cellular Automata*, Arthur W. Burks (ed.). Urbana, Ill.: University of Illinois Press.

Burks, Arthur (1972–1973) 'Logic, computers, and men', *Proceedings and Addresses of the American Philosophical Association* **46**, 39–57.

Burks, Arthur (1975) 'Logic, biology, and automata – Some historical reflections', *International Journal of Man-Machine Studies* **7**, 297–312.

Burks, Arthur (1975) 'Models of deterministic systems', *Mathematical Systems Theory* **8**, 295–308.

Burks, Arthur (1977) *Chance, Cause, Reason – An Inquiry into the Nature of Scientific Evidence*, Chicago: University of Chicago Press.

Burks, Arthur (1979) 'Computer science and philosophy', *Current Research in Philosphy of Science*, pp. 399–420. Peter Asquith and Henry Kyburg (eds.), East Lansing, MI: Philosophy of Science Association.

Burks, Arthur (1980) 'Enumerative induction versus eliminative induction', *Applications of Inductive Logic*, pp. 172–189. L. Jonathan Cohen and Mary Hesse (eds.), Oxford: Oxford University Press.

Burks, Arthur (1984) 'Computers, control, and intentionality', *Science, Computers, and the Information Onslaught*, pp. 29–55. Donald Kerr, Karl Braithwaite, N. Metropolis, David Sharp, and Gian-Carlo Rota (eds.), New York: Academic press.

Burks, Arthur (1986) 'An architectural theory of functional consciousness', *Current Issues in Teleology*, pp. 1–14, Nicholas Rescher (ed.), New York: University Press of America.

Cohen, L. Jonathan (1980) 'What has inductive logic to do with causality?', *Applications of Inductive Logic*, pp. 156–171. L. Jonathan Cohen and Mary Hesse (eds.), Oxford: Oxford University Press.

De Jong, K. A. (1980) 'Adaptive system design – A genetic approach.' *IEEE Transactions: Systems, Man, and Cybernetics* **10**, no. 9.

Forrest, Stephanie (1985) 'A study of parallelism in the classifier system and its application to classifiers in KL-ONE semantic networks', Doctoral thesis, University of Michigan.

Gillies, Andrew (1985) 'Machine learning procedures for generating image domain features', Doctoral thesis, University of Michigan.

Goldberg, David (1983) 'Computer-aided gas pipeline operation using genetic algorithms and rule learning', Doctoral thesis, University of Michigan.

Grefenstette, John J. (editor) (1985) *Proceedings of an International Conference on Genetic Algorithms and Their Applications*, July 24–26, 1985. Carnegia-Mellon University, Pittsburgh, PA.

Holland, John (1975) *Adaptation in Natural and Artificial Systems – An Introductory Analysis with Applications to Biology, Control, and Artificial Intelligence*, Ann Arbor: University of Michigan Press.

Holland, John (1976) 'Adaptation', *Progress in Theoretical Biology* **4**, 263–296. R. Rosen and F. M. Snell (eds.). New York: Academic Press.

Holland, John (1980) 'Adaptive algorithms for discovering and using general patterns in growing knowledge bases.' *International Journal of Policy Analysis and Information Sciences* **2**, 217–240.

Holland, John (1984) 'Genetic algorithms and adaptation', *Adaptive Control of Ill-Defined Systems*, pp. 317–333. O. G. Selfridge, E. L. Rissland, and M. A. Arbib (eds.), Plenum Press.

Holland, John (1983) 'Escaping brittleness: The probabilities of general-purpose learning algorithms applied to parallel rule-based systems', in *Machine Learning II*. R. S. Michalski, J. G. Carbonell, and T. M. Mitchell (eds.) (Extended abstract published in the preprints for the conference, Dept. of Computer Science, University of Illinois, Urbana, Ill.)

Holland, John (1985) 'Properties of the bucket-brigade algorithm', in *Proceedings of an International Conference on Genetic Algorithms and Their Applications*. John J. Grefenstette (ed.), July 24–26. Carnegie-Mellon University, Pittsburgh, PA.

Holland, John and Arthur Burts (1987) 'Adaptive computing system capabable of learning and discovery'. United States, Patent 4,697,242.

Holland, John and Judith Reitman (1978) 'Cognitive systems based on adaptive algorithms.' *Pattern-Directed Inference Systems*, pp. 313–329. D. A. Waterman and F. Hayes-Roth (eds.).

Holland, John, Keith Holyoak, Richard Nisbett, and Paul Thagard (1986) *Induction: Processes of Inference, Learning and Discovery*, to be published by MIT Press in 1986.

Kuhn, Thomas (1962) *The Structure of Scientific Revolutions*, Chicago, Ill.: University of Chicago Press.

Peirce, Charles Sanders: *Collected Papers*. Vols. 1–6 edited by Charles Hartshorne and Paul Weiss, 1931–35. Vols. 7–8 edited by Arthur W. Burks, 1958. Harvard University Press.

Salmon, Merrilee, (ed.) (1989) *The Philosophy of Logical Mechanism: Essays on the philosophy of Arthur W. Burks, with replies*, to be published by Reidel.

Savage, L. J. (1984) *The Foundations of Statistics*, New York: John Wiley.

Scheffler, Israel (1982) *Science and Subjectivity*, Indianapolis, IN: Hackett Publishing Company. Second edition.

Skolnick, Michael (1984) 'Automated comparison of image similarities and differences', Doctoral thesis, University of Michigan.

Smith, Stephen (1980) 'A learning system based on genetic algorithms', Doctoral thesis, Computer Science, University of Pittsburgh, PA.

Uchii, Soshichi (1977) 'Induction and causality in a cellular space.' *PSA 1976*, vol. 2, pp. 448–461. Frederick Suppes and Peter Asquith (eds.) East Lansing, MI: Philosophy of Science Association.

PART III

STATISTICS

ISAAC LEVI

FOUR THEMES IN STATISTICAL EXPLANATION

INTRODUCTION

Suppose an unbiased coin is tossed a million times. Suppose further that the coin lands heads exactly half a million times. The result of the million tosses may be described correctly (a) as a half million cases of landing heads among a million cases of landing heads or tails, (b) as approximately a half million cases of landing heads (or as between 490 000 and 510 000 cases of landing heads) among a million cases of landing heads or tails; and (c) as a million cases of landing heads or tails (so that the relative frequency of heads is one of a million and one values of $r10^{-6}$).

We may explain the occurrence of this event under each one of these three nonequivalent descriptions by appealing to the information that the coin is unbiased, that it was tossed a million times and that the tosses are independent. In each case, the explanatory premisses are the same. However, the transition from explanans to explanandum differs depending on which one of these descriptions of the event whose occurrence is to be explained is deployed in the explanandum. Under the description (a), we may show why the occurrence of the event was not surprising. Under description (b), we may show why the occurrence of the event was to have been expected with confidence. Under description (c), we may show why the event was to have been expected for sure.

Contrast this situation with one where the unbiased coin is tossed a million times but lands heads every time. Description (c) from the previous situation correctly describes the outcome as before. But neither (a) nor (b) does. We can correctly describe the outcome as a million cases of landing heads. Call that description (d).

Explanations utilising precisely the same explanatory premisses may be used here as before. Under description (c), we can explain why the event was to have been expected for sure. And although we cannot

William L. Harper and Brian Skyrms (eds.), Causation in Decision, Belief Change, and Statistics, II, 195–222.

explain why the event was to have been expected for sure, why it was to have been expected with some confidence or why it was not surprising, when described in manner (d), we can explain how, under that description, it was possible for it to have occurred.

The four explanatory themes listed are not restricted to dull textbook applications like coin tossing. Consider a statistical mechanical system in equilibrium at a given energy level. Assuming that the probability distribution over the phase space is the microcanonical distribution, we may consider explaining the frequency distribution of particles of the system in a suitable partition of the "mu-space" (Ehrenfest and Ehrenfest, 1959, p. 26) representing the phase space of a single particle. Just as it is possible for the coin to land heads a million times in a million tosses, it is possible for all of the frequency to be concentrated in a single "cell" in the partitioned mu-space. And the information given about the structure of the statistical mechanical system can be used explain how this is possible. But the event would be very surprising indeed. And it is not to be expected so described. On the other hand, if the frequency distribution conformed to the Maxwell–Boltzman distribution it would be unsurprising so described and the explanatory information could be used to show this. Furthermore, if the frequency distribution were described as approximately Maxwell–Boltzmann distribution it would be unsurprising so described and the explanatory information could be used to show this. Furthermore, if the frequency distribution were described as approximately Maxwell–Boltzmann, the same information could be used to show why the occurrence of this result was to have been expected with con-

I do not mean to suggest that all explanations involving statistical or probabilistic ideas belong to one of the four types listed. Sometimes, for example, statistical claims are the objects of explanation rather than or as well as featuring in the explanatory premisses. In this discussion, however, I mean to focus on the invocation of premisses making claims about chances or statistical probabilities to explain the occurrence of events. These premisses are the sort identified by Hempel (1964, 381–412) as the explanantia of "inductive statistical explanations". In my opinion, much of the recent perplexity concerning the status of inductive statistical explanation arises from the illicit conflation of four different styles of explanation appealing to the same explanatory premisses to reach different explananda.

EXPLANATION BY DISPOSITION

Suppose we are asked why coin *a* landed heads up at time *t*. To say that it was tossed onto a smooth surface shortly before *t* fails to explain causally why the event occurred under the description of that event as a landing heads up at *t*. But allusion to the coin's being tossed shortly before *t* does explain and, indeed, causally explain why the coin landed heads or tails up at *t* (rather than fail to do so). It is not too difficult to see that we can cash this causal explanation in for an argument with the format of a deductive nomological explanation. Two additional premisses are needed:

(1) Coin *a* has a sure fire disposition *D* to land heads up or tails up on a toss onto a smooth surface.

(2) For every *x* and any time *t*, if *x* has *D* at *t* and is tossed onto a smooth surface at *t*, *x* lands heads up or tails up immediately afterwards.

(2) is a general lawlike statement which supports conditionals. The argument is deductively valid and depends for its validity on the use of this lawlike statement. Surely we have all the earmarks of a deductive nomological explanation.

To be sure, an important qualification needs to be made. (2), though lawlike, is a stopgap principle for purposes of explanation. As a rule, it would be regarded as desirable to find some microstructural or other reduction of the predicate *D*. Or if such reduction is not demanded, it is desirable that *D* be integrated into a larger theoretical framework through other lawlike principles in which it occurs nonvacuously additional to the "reduction sentence" (2). Thus, the disposition predicts functions as a "place holder" for purposes of explanation. (Levi and Morgenbesser, 1964.)

Still the place holder status of disposition predicates does not imply that sentences like (2) fail to be statement making or that "*D*" fails to be true or false of coins and other entities. In this respect, one can be as realist as one likes about dispositions. Assigning place holder status imposes a demand on research programs for further explanation; but the explanation would not be a stopgap explanation if (2) could not qualify as a stopgap lawlike premiss of the explanation.[1]

In this qualified sense, we may give an explanation of why the coin landed heads or tails by noting that it was tossed onto a smooth surface. Moreover, even though I lack an analysis of the difference between causal and noncausal deductive nomological explanations, this explanation is as much a causal explanation as is one which explains why a glass broke by noting that it was dropped onto a hard surface, that it was fragile and that fragile things break when dropped.

If the notion of causality plays any serious role in the natural or social sciences or in practical deliberation, it is under the guise of causal explanation that it becomes important. The obsession with singular causation seems misplaced. Still if there is any sense to be made of the notion of singular causation, Davidson's approach is preferable to the diverse conditional analyses (including probabilistic analyses) which have been proposed for our delectation (Davidson, 1967). If coin *a* happens to land heads up after the toss at *t* and I find out, I know that the landing heads up at *t* is identical with the landing heads up or tails up at *t*. Although I cannot causally explain why the coin landed heads up at *t* by noting that it was tossed onto a smooth surface, I may still cite the tossing as the cause of the landing heads up. (See, Steiner, 1983, p. 121.)

There is a more fundamental point lurking behind these remarks. I identified four distinct styles of explanation of events invoking appeal to statistical premisses and information about the occurrence of trial events. The sorts of deductive nomological explanation by disposition I have been considering here are explanations which show why the explanandum was to have been expected for sure. The fundamental point is this: Each of the other types of explanation (explanation why an event was to have been expected with confidence, why an event is not surprising and how it is possible) are successful only if a deductive explanation by disposition shows why the event (under a suitable description) was to have been expected for sure. In this sense, all statistical explanation presupposes deductive nomological explanation and, in those contexts where statistical explanation might be considered causal, it is because the deductive nomological explanation is itself causal.

The point applies quite generally. Statistical explanation of events invokes premisses concerning statistical or chance probability distributions. Such probability distributions characterize an object, system or experimental arrangement of the sort Hacking (1965) called a "chance set up". This characterization utilizes a representation involving three components: (i) a description of a type of experiment, (ii) a "sample space" together with an associated field of "events" and (iii) a probability measure over the field.

The descriptions of the experiment and the events in the field associated with the sets of points in the sample space are intended to represent event types. The points and sets of points in the sample space are surrogates for predicates of event tokens. In our coin example, the experiment description is "is a tossing of coin x" and the sample space has as its two points representatives for "is a landing heads up" and "is a landing tails up".

In order to give structure to the dependencies between the kind of trial, the sample space and the probability measure, various postulates and epistemic constraints need to be introduced. One important but weak postulate is that a chance predicate ascribes a sure fire disposition for any chance set up of which it is true to realize one of the points in the sample space on an experiment of the appropriate type. But this means that the ascription of the chance predicate to a chance set up presupposes that the set up has such a sure fire disposition so that, given that an experiment of the appropriate type has been carried out, it is to be expected for sure that the outcome described as a realization of some point or other in the sample space will occur.

The objection might be raised that chance predicates do not always contain reference to a kind of experiment or other initiating event. In a Poisson process, for example, the experiment may be characterized as an elapse of time; and elapses of time are not kinds of events and, hence, not kinds of experiments which may be cited in causal explanation.

To the extent that this is true, the objection shows that the explanation by disposition grounded on the chance assumption may not be causal. Even so, the explanation by disposition remains deductive nomological and shows why the explanandum event described as the realization of some point or other in the sample space was to have been expected for sure. Moreover, in many applications of a Poisson model, the explanation by disposition may reasonably be taken to be causal. In the paradigmatic case of a telephone exchange, the kind of experiment which sets up the Poisson process is not an elapse of time but the construction of the telephone exchange and the placing of that exchange in operation for a certain period of time. Relative to this kind of experiment, the sample space consists of the possible sequences of dated calls during the period which the exchange is running. The operation of the exchange causally explains why some sequence or other in the sample space is realized.

By way of contrast, we do not manufacture radioactive elements and place them in operation. At best, we hook the system up to some emission detector; but then the sample space is the possible number of emissions registered by the detector during an interval of specified duration and not the possible number of emissions during that interval of time. Nonetheless, we may explain deductive nomologically by disposition why some sequence or other of dated emissions in the sample space was realized even though the explanation is not causal.

Another context where explanation by disposition may arise without being causal is statistical mechanics. Suppose we have a gas described macroscopically as being in a container of given dimensions and being at a given energy level over a specified period of time. The sample space of possible trajectories of the system in phase space can then be specified. Such a system has a sure fire disposition to realize some trajectory in the sample space; but, in the absence of any account of how the system is produced and set in motion, we have no characterization of an experiment relative to which the statistical by disposition can be regarded as causal. To be sure, sometimes such systems are "prepared" and then the explanation by disposition may be regarded as causal.

Whether such explanations are causal or not, explanations of the occurrence of events appealing to assumptions about chances all presuppose the availability of an explanation of the occurrence of the same event (under a suitable description as the realization of some point or other in the sample space) by sure fire disposition which shows why the occurrence of the event was to have been expected for sure. This is so regardless of whether the explanation purports to show how the event is to be possible, why it is to be expected with confidence or why it is not surprising.

Furthermore, even though not all such explanations by disposition are causal, the only causal explanations involved in the use of statistical premisses in the explanations of events are such deductive nomological explanations by sure fire disposition.

If this is right, there is not the slightest reason why we should countenance any type of causal explanation other than deductive nomological explanation. Even when statistical premisses are invoked in causal explanation, the causality is not probabilistic.

EXPLANATION HOW POSSIBLE

In characterizing a sample space relative to a kind of experiment, we need to do more than assume that the system has a sure fire disposition to realize some point or other in the sample space. For each set A of points in the field, it is necessary also to assume that the system has an ability to realize a point in A on an implementation of the given kind of experiment. To have an ability to respond in manner R on a trial of type S is the dual of having the sure fire disposition to respond in manner R on a trial of type S. The system having the ability does so if and only if it lacks the sure fire disposition to fail to R on a trial of kind S. Hence, the description of the sample space implies information about abilities as well as a sure fire disposition and this information may also be used in an explanatory way.

For example, even though we cannot invoke a deductive nomological explanation from the statistical premisses concerning coin a to show why it was to have been expected that a landed heads up under that description of the event as a landing heads up, we can show why it was not to have been ruled out that the coin landed heads up.

The ability of a system to behave in a given way relative to a certain type of trial (the possibility for the system to behave in that way on that kind of trial) is an objective feature of the system. Like statistical probability or chance, possibility for applies to event types and is relative to experiments. Coin a has the ability to land heads up on a toss (even if it is never tossed) but at the same time it lacks the ability to land heads up on a toss described as placing the coin in a mechanical state which leads by the laws of mechanics to tails.

Such objective possibility contrasts with subjective or epistemic possibility (possibility that) which applies to sentences. A hypothesis h is a serious or subjective possibility according to agent X at time t if and only if h is consistent with X's knowledge at t.

To *explain* how it is possible *that* the coin landed heads at t requires showing that the explanandum is consistent with a given body of information.

It is not enough, however, to show that the hypothesis that the coin landed heads is a serious possibility. To be explanatory, the judgement of serious possibility must be relative to an explanatorily appropriate body of information. In particular, the judgement of serious possibility must be grounded on (i) the information that the coin has the ability to land heads on a toss (that it was possible for the coin to land heads on a toss) and (ii) the information that the coin was tossed.

To see the importance of requiring information of kind (i) and (ii) in explanation how possible, consider the following three variant scenarios concerning coin *a*:

Case 1: Whether the coin is tossed at time *t* or not is unknown. That the coin landed heads at *t* is a serious possibility; but we cannot explain how it is possible by appealing to the coin's abilities and the information that it was tossed.

Case 2: It is known that the coin was tossed at *t*. However, it is unsettled as to whether the coin has the ability to land on its edge on a toss. Hence, it is seriously possible that the coin lands on its edge at *t*; but the premisses cannot give us an explanation of how the coin's landing on its edge is possible.

Case 3: It is known that the coin was tossed at *t*. It is also known that the coin is able to land heads on being tossed. Provided that there is no further information available entailing that the coin did not land heads up, it is seriously possible that the coin landed heads up just as in case 1. However, in contrast to case 1, the judgement of serious possibility here is grounded on knowledge of the coin's abilities. We can explain how it was possible that the coin landed heads up.

Thus, we have a second type of explanatory argument which invokes statistical premisses but makes no essential use of the probabilistic information. We may explain why it was possible that the fair coin tossed a million times which landed heads up every time did so under that description (description d) by noting that landing heads up a million times is a point in the sample space relative to an experiment in which the coin is tossed a million times. Similarly one can explain why it is possible that all the molecules in the gas are located in one cell of the mu-space.

Of course, one can explain the occurrence of any point in a given sample space in this way. For those who (rightly I believe) think that the probabilistic information in statistical premisses sometimes has a significant role to play in the explanation of events, there should be styles of statistical explanation of events other than explanation how possible and explanation via disposition.

Before proceeding to explanatory arguments which appeal to probabilistic information in a more substantial way, one objection needs to be faced.

As I have characterized it, explanation how possible shows that a certain statement (the explanandum) is a serious possibility relative to a body of knowledge satisfying certain requirements. This way of looking at the matter works well enough in predictive contexts where the agent knows that the set up has the abilities in question but does not know the outcome of the experiment. Then every hypothesis specifying that a point in the sample space is realized is a serious possibility – i.e., is consistent with the body of knowledge.

In the context of explanation, however, the agent knows which point in the sample space is realized. The problem is that given what is known, we can do more than establish serious possibility. We can establish certainty.

The point of explanation how possible is not, however, to demonstrate what is already known but to show what should be concluded from the explanatory premisses even if the agent were not to know the result of the experiment. Just as deductive nomological explanation shows why the explanandum was to have been expected for sure even if one did not already know that it held, explanation how possible shows why the explanandum was not to have been ruled out even if one did not already know that it held.

Thus, the body of knowledge endorsed by the investigator seeking an explanation does not meet the conditions required for explanation how possible. Hence, the epistemic relativity of explanation how possible is to a standard for serious possibility different from the one endorsed by the agent. The question is: How is this alternative standard to be characterized?[2]

I suggest that the standard is to be derived from the standard currently endorsed by the investigator through an exercise in hypothetical belief revision.

Suppose that, to begin with, the agent knows that the coin has been tossed and has landed heads. He also knows that the sample space for tossing of the coin consists of the possible outcomes of landing heads up and landing tails up. Hence, the agent knows that whenever the coin is tossed it lands either heads up or tails up. He also knows that the coin has the ability to land heads up on a toss and also to land tails up on a toss. Let us call the agent's initial body of knowledge *K*.

The agent might contemplate the consequences of removing from *K* the information that the coin has landed heads up or tails up. Because the resultant corpus must be deductively closed, he needs to give up something else. According to the account of contraction of a corpus of knowledge I have proposed elsewhere (Levi, 1976, 29–30; 1977b, 168–169; 1980a, 58–63; 1983, 122–125), the agent should adopt a contraction strategy which minimizes loss of informational value. In this example, less is lost by giving up the claim that the coin is tossed than that it has the sure fire disposition to land heads or tails on a toss and that whatever has such a disposition lands heads or tails whenever tossed.

Having contracted from *K* to *K*∗, the question of the outcome of tossing has been converted to a question of prediction. Adding the claim that the coin was tossed to *K*∗ and taking the deductive consequences to form the expansion *K*∗∗ of *K*∗, it follows logically that the coin lands heads or tails. We have shown why it was to have been expected for sure that this would happen. Furthermore, relative to *K*∗∗, it is a serious possibility that the coin lands heads up (and also that it lands tails up).

K∗∗ is not, of course, the agent's body of knowledge *K*. But it is derivable from *K* by an appropriate series of transformations. I suggest that it is a serious possibility relative to *K*∗∗ that is derived from information about objective possibility in explanation how possible.

The procedure for obtaining the corpus *K*∗∗ from *K* entails first removing from the doctrine all information implying that a point in the sample space is realized and doing so in a manner which otherwise minimizes loss of informational value. Once this is done, add to the resultant standard *K*∗ all assumptions lacking for the putative explanation to form *K*∗∗. Assessments of certainty and serious possibility for purposes of deductive nomological explanation via disposition and explanation how possibility are relative to *K*∗∗.

DIRECT INFERENCE

Relative to the corpus *K*∗∗ described above, it is seriously possible that the coin lands heads up and it is also seriously possible that it lands tails up. The assessments of serious possibility are derivable from information that the coin has the ability to land heads up (tails up) on a toss and that the coin was tossed on the specific occasion *t*. To be able to obtain the assessments of serious possibility, it is crucial that the agent not have in $K*$ additional information about the kind of experiment the tossing at t is which rules out the hypothesis that the coin lands heads up (tails up) as a serious possibility.

This epistemic feature of argument from knowledge of objective possibility to subjective (i.e., serious) possibility is paralleled in "direct inference" from knowledge of objective probability (chance) to subjective (credal) probability. If the agent has in $K*$ the information that the chance of obtaining heads on a toss is 0.5 and likewise for tails and he knows that the coin is tossed at t (also in $K**$), then relative to $K**$ the credal probability which should be assigned the hypothesis that the coin lands heads up (tails up) should be 0.5 – provided the agent does not have additional information in $K**$ which frustrates the assignment – e.g., information to the effect that the coin was tossed with the aid of a device which favors tails so that the chance of tails on a toss with that device is 0.9 and of heads is 0.1.

The question of specifying the kind of information about the trial which does and does not frustrate the direct inference is a matter of some controversy. The problematic cases concern situations where the agent knows additional information about the kind of trial (e.g., that it is tossed with a special device) but does not know what the chances of heads (tails) are on a toss of that kind.

Kyburg (1961 and 1974) takes the position that when such additional information allows a range of hypotheses about the chances to be serious possibilities which spans 0.5, one should assign a credal probability of 0.5 to the hypothesis of heads (tails) as if one knew only that the coin was tossed and knew nothing additional concerning how it was tossed. I take the position that the additional information is relevant and prevents assigning credal probability of 0.5 as obligatory (Levi, 1977a; 1980a, 12.4 and 16.2–16.4 and 16.6). On my view, one should always use all the information one has about the kind of trial and if chances are indeterminate relative to trials so described, direct inference does not specify the credal probability except that it fall in the range of indeterminacy.

I have discussed this issue elsewhere and do not intend to rehearse the points under dispute here. For the purpose of this discussion, I shall dogmatically affirm my view of the matter. Armed with this approach, it will then prove possible to discuss the two explanatory themes involving statistical assumptions which make essential use of probabilistic information: explanation why the occurrence of an event was not surprising and why the occurrence of an event was to have been expected with confidence.

Notice that the condition on the information about the kind of trial which I advocate for purposes of direct inference is tantamount to a condition of epistemic homogeneity. One may ignore extra information about a trial if it is known to be stochastically irrelevant.

Of course, in explanatory contexts, the outcome of the experiment to be explained is already known and is in *K*. However, as suggested in connection with explanation how possible, *K* should be transformed to *K*** by first contracting to *K** through removal of all information entailing that a point in the sample space is realized (with minimal loss of informational value) and then expanding through the addition of the information that the experiment is conducted.

Thus, in the statistical mechanical example, we are required to remove the information that the system has followed a trajectory on the energy surface and, hence, the information that it was prepared to be at that energy level. This gives us *K**. *K*** is obtained by adding the information that the system was prepared to be at the given energy level.

Observe that we may then use the information that the objective chance distribution over the phase space is the microcanonical distribution to derive a credal probability distributionover all possible trajectories. That is because the information about how the system was prepared is epistemically homogeneous. It is not, however, objectively homogeneous and, moreover, is known not to be objectively homogeneous in Salmon's sense (1984, ch. 3) even relative to the transformed doctrine *K***.

ACCEPTANCE

When combined with information about the payoff structures, credal probabilities derived from knowledge of chances suffices for the purpose of placing bets and, more generally, for determining expected values. But assigning credal probabilities to hypotheses does not suffice for determining whether these hypotheses should be added to the information constituting the initial corpus. In statistical prediction and explanation purporting to show why a hypothesis is not surprising and to show why it is to be expected with confidence, we seek to determine which of a list of potential answers to a given question are to be ruled out by induction and which are not to be eliminated on the basis of the information contained in the explanatory or predictive assumptions.

The truth of h is not surprising relative to a body of information if and only if the acceptance of $\sim h$ via inductive inference from that information is not legitimate. The truth of h is to be expected relative to that body of information if and only if its inductive acceptance is legitimate.[3]

Thus, given the information that coin a is tossed a million times, is unbiased and tosses are stochastically independent, it is surprising for the coin to land heads up every time. That is to say, a reasonable agent might infer via induction that such an event will not occur. On the other hand, the hypothesis that the coin lands up exactly 500 000 times is not surprising. A reasonable agent has no warrant for concluding via induction that this will not occur. Finally, a reasonable agent should be prepared to accept via induction that between 490 000 and 510 000 tosses will yield heads up.

These claims need qualification. Several contextual parameters additional to the information available in the initial corpus (the available evidence) need to be specified in order to determine the legitimacy of such inferences (Levi, 1967a, 1967b, 1979, 1980a). When I claim that the verdicts described above are "reasonable", I am assuming that the values of these parameters are specified in what might plausibly be taken to be standard ways.

For example, the legitimacy of inductive expansion depends on the list of potential answers to the question being asked. Such a list can be given by starting with an "ultimate partition" and considering all hypotheses equivalent given the initial body of information with Boolean combinations of elements of the partition (Levi, 1967a). In the coin example, most of the time our interest is in the prediction or explanation of relative frequencies of heads and tails and, hence, the ultimate partition should consist of all million and one hypotheses specifying an exact relative frequency for heads and for tails. Relative to this partition, one can generate all hypotheses which are disjunctions of elements of the partition.

Of course, inductive inferences might be assessed relative to another ultimate partition. Insofar as the choice of an ultimate partition makes a difference to the legitimacy of inductive inferences, it reflects a difference in an aspect of the inquiring agent's demands for information.

Another contextual feature reflecting a different aspect of the demands for information is the way in which the potential answers generated by the ultimate partition are assessed with respect to informational value. I have proposed representing such an evaluation by a probability measure over the boolean algebra generated by the elements of the ultimate partition (Levi, 1967b). Often, though not always, the information determining probability assigned to each element of an ultimate partition will be the same. I take it to be plausible in the case of the coin tossed a million times to suppose that equal informational value will be assigned to each hypothesis of the form "The relative frequency of heads is $r10^{-6}$." (r is an integer from 0 to 1 000 000.) That informational value would equal 1/(1 000 001).

Notice that if we had chosen all possible sequences of a million heads and tails as our ultimate partition, we could still assign equal informational value to the hypotheses about relative frequencies by using an information determining probability (an M-function) formally identically with Carnap's $c*$. The elements of this ultimate partition do not receive equal informational value. However, as before hypotheses about precise relative frequencies do.

For many purposes, it makes little difference whether we proceed in this fashion or adopt the million and one frequency hypotheses as our ultimate partition.

It should be emphasized that the information determining probability function (the M-function) is not to be confused with the current credal probability or with some prior credal probability from which the current credal probability is derivable via conditionalization. Nor is it to be confused with objective chance. For reasons I shall not consider here, I accept the view that informational value increases with decrease in some probability measure or other. Indeed, I find it convenient to use as the measure of informational value the magnitude $1 - M(x)$. But the M-function used here represents a demand for information on the part of the agent and neither the agent's current state of belief nor some prior state. $1 - M(x)$ represents an epistemic value or utility expressing the desideratum of acquiring new information over and above that entailed by the initial unexpanded body of information through adding x and all the consequences of x and the unexpanded doctrine regardless of whether x is true or false (Levi, 1967b).

According to the account of inductive inference I favor, in inductive expansion one should accept the potential answer which furnishes the best trade off between the informational value obtained and risk of error incurred (Levi, 1967a, 1967b). In order to assess this, we need to be able to strike a balance between the importance of risking error and of obtaining new information. This balance is represented by an index q which increases with the boldness of the investigator (decreases with his caution). The greater q is the greater the relative importance attached to informational value as compared with risk of error. Bypassing technical details, this way of looking at matters suggests that the best potential answer to a given question given the initial doctrine K, the credal state represented by credal probability Q, the ultimate partition, the information determining M-function and the index q is that potential answer g for which the expected epistemic utility $Q(g) - qM(g)$ is a maximum.

This recommendation together with a rule for breaking ties (Levi, 1967a) recommending the choice of the weakest conclusion which bears maximum expected utility leads to the prescription that all and only elements of the ultimate partition for which $Q(h) - qM(h)$ is negative be rejected.

Given the further condition that q is restricted to values between 0 and 1, there will be some elements of U which are not rejected for any value of q in the permitted range. These are completely unsurprising. Suppose that element h of the ultimate partition fails to be rejected for values of q below $q(h)$ but is rejected for values greater than $q(h)$. We may say that h is surprising to degree $1 - q(h) = d(h)$. If h and g are both elements of the ultimate partition, $d(h \vee g) = \min(d(h), d(g))$. (Levi, 1967a. See also, Levi, 1984a, ch. 13.)

In our example, direct inference warrants assigning the hypothesis that every toss in the million lands heads the value $(1/2)^{1\,000\,000}$. This is very much less than the value 1/(1 000 001) which is the rejection level if $q = 1$. Hence, this hypothesis bears a positive degree of surprise and, indeed, a substantial one.

On the other hand, the hypothesis that the coin lands heads exactly a half million times bears a probability of

$$\{(1\,000\,000!)/[500\,000!\ 500\,000!]\}(1/2)^{1\,000\,000}$$

which, though very small, is substantially higher than 1/(1 000 001). No matter what value for q between 0 and 1 is adopted, this hypothesis avoids rejection. It is totally unsurprising.

To say this, however, is not to claim that one is justified relative to the initial body of information in adding this hypothesis to that body of information via induction. Even when $q = 1$ so that the agent is maximally bold or minimally cautious, other specific hypotheses concerning the exact relative frequency would go unrejected. There are other perfectly unsurprising elements of the ultimate partition. The disjunction of all of these will itself be perfectly unsurprising. Furthermore, the disjunction of all the remaining elements of the ultimate partition will be positively surprising because even the least surprising of its disjuncts is positively surprising.

This means, however, that the disjunction of all the perfectly unsurprising elements of the ultimate partition is legitimately accepted if $q = 1$ and is, in this sense, to be believed with some positive degree of confidence. To be sure, if q is less then 1, one might not add this disjunction into the background. But by adding more disjuncts, a new proposition is added at the level of boldness $q*$. We may say that this proposition is believed with a degree of confidence somewhat greater than the one accepted when $q = 1$.

It is in this sense that we can show that the hypothesis that the coin lands heads with a relative frequency between 490 000 and 510 000 is to be expected with a positive degree of confidence.

Thus, we are now in a position to characterize the way in which one explains why it is not surprising that the coin lands heads up exactly 500 000 even though we cannot explain why it was to be expected that this event so described occurs. We begin with the corpus K where it is known that this event occurs and delete the information that a point in the sample space has occurred to form $K*$. We then add the information that the coin is tossed a million times. To obtain an explanation, we need an ultimate partition and shall suppose that it is the set of hypotheses about relative frequency and that each element of the ultimate partition is given equal M-value. The expectation determining credal probability is derived via direct inference from the information about the chances and trials given in $K**$. Given this information, we can determine which elements of the ultimate partition are to be rejected for various value of q and in this way show that the hypothesis that the coin lands heads exactly a half million times is completely unsurprising though not to be expected.

An entirely parallel process can be undertaken when the coin is known to have landed heads every time. But here we cannot show that the outcome so described is unsurprising except at an extremely high degree of caution. We can, however, show how it is possible.

ACCEPTANCE AND PROBABILITY

Landing heads up every time in a million tosses and landing heads up exactly half a million times in a million tosses are both extremely improbable. Yet, explanation how possible is available in both cases. Indeed, landing heads up exactly half a million times can even be shown to be unsurprising.

Probabilists have often thought of a requirement to accept hypotheses with a probability higher than some threshold value (which is tantamount to rejecting their negations when their probabilities go below some threshold) as a fundamental desideratum of any account of inductive acceptance. Such a requirement need not be satisfied according to the criteria for acceptance I have suggested. One can entertain situations where the elements of the ultimate partition which go unrejected when $q = 1$ (and even when it is less than 1) have a total probability which remains below 0.5. Hence, when inductive inferences are used in explanatory efforts to show why an event was to have been expected, it is not necessary that such efforts establish a high probability.

However, when degrees of belief are construed in Shackle's way, one can say that such explanatory efforts must show why the hypothesis that a certain event occurs ought to be believed or accepted at a given level of confidence in Shackle's sense (Levi, 1967a, and Cohen, 1977). In this sense, to say that a hypothesis is to be accepted if its degree of confidence of acceptance or degree of belief is higher than some threshold r is to say that the level of caution q for rejecting elements of the ultimate partition is set at $1-r$ and at that level the hypothesis in question is accepted.

I conjecture that the reason why so many people find high probability rules attractive is a failure to distinguish carefully enough between measures of degrees of belief or of support which are expectation determining measures such as credal probability and measures, like Shackle measures, whose values are required to reach a satisfactory level before hypotheses may be accepted.

The confusion of credal probability (which functions in the determination of expected value) with satisficing measures of support has been compounded with the conflation of expectation determining measures with indices (also called measures of support, confirmation, probability or likelihood) whose values are to be maximized in deciding what to accept via induction from the available evidence and background knowledge. Clearly, however, probability cannot be both a satisficing and maximizing measure of support unless one insists that the only hypotheses to be accepted relative to the available evidence bear probability 1 – a point of view which, in effect, denies the legitimacy of inductive inference altogether.

Nor does it make sense to choose either a satisficing or a maximizing role for credal probability. To do the former is to favor violations of the requirement that a standard for serious possibility be deductively closed. To do the latter is to recommend once more that only items bearing maximum probability be accepted.

According to the proposals I have sketched here, the index to be maximized in deciding which of a set of potential answers to a given question to answer is the expected epistemic utility measure. This is the index which assigns the value $Q(h) - qM(h)$ to accepting h as strongest via induction – i.e., adding h and all the deductive consequences of h and K to the initial corpus K. The index to be satisfied is the measure of degrees of belief or confidence of acceptance in the Shackle-like sense.

Both of these indices are related in a definite manner to credal or expectation determining probability measures (as well as to information determining probability measures). However, neither index is reducible to expectation determining probability as probabilists tend to require.[4]

BOOKKEEPING

Suppose we proceed to reject elements of some ultimate partition relative to initial body of knowledge K, information determining probability M, credal probability Q and index of boldness q. The result will be an expansion to a new corpus K_1 in which only a subset of the elements of the original ultimate partition survive rejection. These survivors will be assigned new credal probability Q' and new information determining probability M' by conditionalization so that the rejection procedure may be reapplied. The process may be reiterated until no further elements are rejected. The outcome of such bookkeeping can explored. For example, when $q = 1$ and the *M*-function assigns equal value to all elements of the ultimate partition, the only elements of *U* which survive rejection via bookkeeping are those which bear maximum credal probability relative to the initial corpus. Thus, in the bookkeeping sense, it can be shown to be expected that the coin tossed a million times will land heads up exactly a half million times if q is set at 1 and not merely that it is unsurprising. However, if q is less than 1, this will no longer be true.

STATISTICAL EXPLANATION

If one adopts this point of view (it is the point of view I have advocated since (Levi, 1967a)), it is unnecessary to derive a high probability for an explanandum in order to show that it is not surprising or to show that it was to have been expected. Examples like the much discussed case of paresis present no difficulties. Granted that contracting paresis is not to be expected in tertiary stage untreated syphilis, it remains more than a possibility. It is unsurprising. We can explain why it was to be expected that the flat tire did not spontaneously inflate by appeal to considerations of a statistical mechanical nature. Macroscopically detectable entropy reversals would be positively surprising in systems with large numbers of particles (in contrast to the Brownian motion). On the other hand, slight and undetectable reversals would not be surprising.

Consider a gas constituted of *N* (*N* large) similar particles confined to a container of a given volume and held at a fixed energy level plus or minus a small variation.

Divide the phase space for a single particle (mu-space) into finitely many small cells of equal dimension and define the macroscopic condition of the gas as the distribution of the particles in these cells. Such a distribution is determined by the mechanical state of the total *N*-particle system as represented by a point in the phase space for that system (gamma space). This distribution, in turn, is related to the phenomenological descriptions of the system in terms of volume, pressure, energy, temperature, and entropy at least in suitable conditions of thermodynamic equilibrium.

In classical statistical mechanics, it is taken for granted that the frequency distribution of particles in the cells into which gamma space is partitioned will be satisfactorily approximated by the Maxwell–

Boltzmann distribution in thermodynamic equilibrium (at least if N and the numbers of particles in most of the cells is large). When this is so, one can proceed to argue for the standard equilibrium conditions familiar for thermodynamic systems of ideal gases.

Our interest here is with the explanation of the Maxwell–Boltzmann distribution from explanation mechanical considerations. Such explanation (or prediction) is to be contrasted with attitudes towards a distribution of the N particles in mu space which locates all the particles in a single cell.

The familiar argument runs as follows: When the gas is in statistical equilibrium (never mind, for our purposes, how it got there) at a fixed energy level, the distribution of chances over phase space is the microcanonical distribution. If the gas is in thermal equilibrium with some heat reservoir, the appropriate distribution is the canonical distribution. In either case, the Maxwell–Boltzmann distribution among the cells is much more probable than other frequency distributions. The distribution locating all the particles in a single cell is extremely improbable and would be considered extremely surprising.

Sense may be made out of this if one considers the set of hypotheses specifying frequency distributions in the cells in mu space to be be the ultimate partition. This is formally analogous to using Carnapian structure descriptions as the ultimate partition or, in the case of the coin, of using the $n+1$ hypotheses specifying possible relative frequencies for heads in n tosses. A frequency distribution specifying that all the particles are in one cell will have such small probability that it would be rejected by all but the virtually sceptical – i.e., relative to all values of the index of caution q except those very close to 0. Hence, the degree of confidence of rejection (or potential surprise) is not only positive but extremely close to the maximum value of 1.

By way of contrast, the Maxwell–Boltzmann distribution bears the highest probability of all elements of the ultimate partition. If $q = 1$ and bookkeeping is deployed, one may show that it is to be expected. If q is less than 1 or if one uses the single shot method, it can be shown that this distribution is not surprising just as the prediction of exactly half a million heads was shown to be unsurprising. As in the coin tossing case, the hypothesis cannot be shown to have been expected if q is appreciably less than 1.

Regardless of how one proceeds, if all frequency distributions are assigned an equal M-value, it can be shown that a distribution which exhibits only small deviations from the Maxwell–Boltzmann distribution is to be expected. As in the coin tossing case, this result emerges not because the hypothesis that a small deviation will occur bears high probability (which it does) but because the chance induced probability distribution is unimodal and has a small variance.

It is sometimes held that the Maxwell–Boltzmann distribution should be endorsed because it is the most probable distribution of frequencies among the cells in mu space (for example, Boltzmann and Boltzmann, 1959, p. 50) or, at least, that that distributions very near the Maxwell distribution are to be expected because the probability of this being so is very near certainty (Malament and Zabell, 1980).

Such arguments, it seems to me, extrapolate fallaciously from a feature peculiar to situations where distributions over the ultimate partition are unimodal (with means equalling modes) and where the dispersion is small. This causes all sorts of grief for accounts of inductive inference in general and probabilistic explanation in particular.[5]

Thus, the example of the fair lottery ought to show decisively that high probability cannot be sufficient for acceptance. In a fair lottery, the sensible response is to suspend judgment completely as to the outcome – not to predict that ticket *i* will not be selected for each value of *i*. I am far from subscribing to the principle of insufficient reason according to which, if I am ignorant as to which ticket will be drawn, I should assign equal probability to the alternatives. Still, advocates of the principle are on the right track when they suppose that if the alternatives bear equal probability one should remain ignorant of the outcome of the lottery.

THE CONTEXT DEPENDENCY OF ACCEPTANCE

Of course, on my view, the conclusions to be adopted via induction are sensitive to the choice of a set of potential answers and to an assessment of informational value and this sensitivity applies in contexts of statistical explanation as well as prediction. Some authors appear to find this type of context sensitivity threatening to the objectivity of scientific inference.

Let me recall the worry registered in section 30 of Ehrenfest and Ehrenfest, 1959 and discussed with greater elaboration by Tatiana Ehrenfest in Ehrenfest, 1958.

T. Ehrenfest invites us to consider an urn containing three balls of which two are red and one black.

She raises the question of what to expect on a single trial through sampling by replacement from the urn.

She then asks what to expect on n repetitions where one focuses on relative frequencies of reds and blacks. She calls this problem one of second order with index n.

She then considers a sequence of n_2 n-fold samples with replacement. These are third order problems with index (n, n_2). One can, of course, reiterate further.

Ehrenfest points out that the most probable second order outcome is a relative frequency of 2/3 reds if n is divisible by 3 or one as close as possible to that frequency. The most probable outcome of third order is a sequence of second order outcomes with each second order sequence occurring with a frequency proportional to its probability.

Ehrenfest then alleges that the usual attitude is to expect the most probable outcome of the order of interest of an experiment of this kind. According to the position I have advocated, this will not, in general be true. However, under special conditions it will hold. If the ultimate partition adopted consists of the possible frequencies in the given order with the appropriate index, $q = 1$, each element of the ultimate partition is assigned equal M-value and bookkeeping is deployed, the element of the ultimate partition bearing maximum probability should be accepted. But even $q < 1$ or bookkeeping is not used, the sensitivity to the order of the problem to which Ehrenfest draws our attention remains. Choosing the order of the problem becomes virtualy equivalent to identifying an ultimate partition whose elements receive equal M-value.

Ehrenfest points out that if one focuses on expectations according to first, second and third order questions, one obtains conflicting results. In the first order problem, one would expect to draw a red. Ehrenfest then notes that in a large number of repetitions, it is to be expected that these first order predictions will be false about 1/3 of the time. And if one then proceeds to consider a large sequence of second order sequences, the most probable third order result will be one in which even a sequence of nothing but lacks will occur at least some small proportion of the time.

Ehrenfest does not question, as I would, the criterion of accepting the most probable alternative. She does, however, see quite clearly that in fixing beliefs here one focuses on some specific list of potential answers – in effect, on a given ultimate partition.

In closing her paper, Ehrenfest does not conclude that the choice of an order is entirely a subjective matter. There is a kind of argument one might give for picking one order rather than another:

> One considers a phenomenon of which one knows enough to attribute to it a certain stability. One has reason to believe certain values observed to be statistical averages of certain values. This determines the order and the index of our hypothesis. If the sensitivity of our observation were more acute, if we could distinguish differences of pressures of a gas in microscopic areas, if we were thinking microbes doing physics, it would not be the hypothesis of the second order with index n_1, but a hypothesis of third order with index (n_1,n_2). (1958, p. 391).

Ehrenfest is suggesting that the greater sensitivity of "thinking microbes" would render the macroscophic phenomena which appear stable to us a blooming buzzing confusion. They would, therefore, be inclined to seek stable phenomena as the objects of explanation differently than we do and to regard the demands for stability they make as part of the "objective" case for the type of potential answers they entertain and the way they assess their informational value.

This interesting idea suggests that, at least in some contexts of prediction and explanation, the choice of a list of potential answers may be constrained by considerations internal to the scientific domain involved.

However, even where the choice of an "ultimate partition" and an information determining *M*-function is subject to controversy which cannot be resolved along lines hinted at by Ehrenfest, it is possible, as I have shown in Levi, 1979 to reconstitute inductive arguments by resorting to shared agreements in a way which allows for sensible results.

Thus, the pragmatic and contextual character of statistical explanation showing why events are not surprising or are to be expected does not imply the presence of an undesirable lack of objectivity.

Even those who are reluctant to make inductive inferences in which hypotheses are accepted via induction from initial information cannot avoid totally the need to take epistemic and pragmatic considerations into account. According to the account of direct inference, I favor the requirement of epistemic homogeneity is imposed on the description of the experiment relative to which statistical information is referred.

Kyburg's approach, though different from mine still contains an ineliminable epistemic dimension.

THE EPISTEMIC STATUS OF STATISTICAL EXPLANATION

Coffa (1974) and Salmon (1984, pp. 48–55) have complained about the epistemic character of statistical inductive explanation on Hempel's model and presumably their complaints are intended to apply to the rather different models suggested by Kyburg on the one hand and by me on the other. They allege that those who emphasize the epistemic and pragmatic aspects built into statistical explanations are committed to regarding such explanations as second rate substitutes for more objective deductive nomological explanations.

Recall that the requirement of epistemic homogeneity I favor applies not to epistemic homogeneity relative to the agent's current body of knowledge but relative to a certain transformation of it. One is to remove from the current body of information all assumptions implying that a point in the sample space occurred.

No doubt there is some ambiguity here; for there is often considerable leeway concerning how the sample space is to be specified. A Laplacian demon who knew the position in phase space of a gas at some time and sought to explain its position in phase space at some later time (or, perhaps, to explain some function of the phase) might rest content with removing just the information about the phase points without removing the information that a point on such and such an energy surface in phase space is realized. In that case, the demon's transformed information set will still contain information which should frustrate a direct inference from knowledge of the microcanonical distribution. He could, after all, continue to hold that the phase is very close to the specific value he deletes from his information. Hence, statistical mechanical premisses are not available to him for explanatory purposes – even though he knows them to be true.

Even so, the Laplacian demon might acknowledge that the explanations he would be able to offer would not be applicable to large numbers of other systems similar to the one under study with respect to observable thermodynamic properties. In order to obtain the requisite broad coverage, even the Laplacian demon might want to resort to a statistical explanation. To do so, he would have to delete from his information set the hypothesis that some point or other on the energy surface had been realized. This would place the demon in the hypothetical information set from which he could then expand by adding the putative statistical explanation and nothing else in a way insuring epistemic homogeneity. Relative to this hypothetical information set, the agent would know that the information about the kind of trial is not objectively homogeneous – at least in the context of classical statistical mechanics where the underlying scheme is deterministic.

The point to emphasize is this: Even though the demon has at his disposal information sufficient to furnish a deductive nomological explanation of the events under study, a hypothetical information set is adopted so as to avoid doing so. The reason why the demon who, unlike ourselves, could offer a "deterministic explanation" avoids doing so is that the demon, quite like ourselves, seeks to furnish explanations displaying the kind of generality which statistical mechanical explanations can provide. Coffa and Salmon to the contrary notwithstanding, statistical explanations are not second rate substitutes for deductive nomological explanations – even when one insists on requirements of epistemic homogeneity and other context sensitive conditions.

Columbia University

NOTES

[1] The placeholder view of dispositionality has been criticized on the grounds that when a disposition predicate is replaced by a specification of microstructure, new disposition predicates are introduced into the characterization of the "basis". Dispositionality is allegedly ineliminable.

The kernel of truth in this observation constitutes no objection to the placeholder view. When a disposition predicate is given a basis, the explanatory value is enhanced due to the integration of the disposition predicate into a theoretical structure – not due to the avoidance of other disposition predicates. The placeholder view does not see the defect in placeholding status of disposition terms lying in the absence of an eliminative reduction to a basis free of disposition terms but in the absence of integration of such terms into an explanatorily adequate system. The conditions for adequate integration are vague and vary with the program for explanation being favored (Levi and Morgenbesser, 1964). But whatever they are, they need not imply full replaceability of the placeholder. They might require that a more powerful system of axioms has been endorsed which lend further structure to items of which the disposition term is predicated. Such systematisation might convert the term from the status of a placeholder to an explanatorily adequate theoretical primitive in the context of the particular discipline in which it is used.

The possibility of integration without elimination becomes even more critical in the case of objective probability. There is a tradition in statistical mechanics which seeks to dispense with the use of objective, statistical probability or, if not that, to explain it. But even the mightiest efforts of advocates of the importance of the concept of ergodicity do not claim that with a certain type of assumption of ergodicity it becomes possible to derive the microcanonical distribution as the sole stationary distribution over the given energy surface without any probability assumptions. Not only must it be assumed that the probability measure is stationary and countably additive but that it be absolutely continuous in Lebesgue measure. Some authors seem to conclude that insofar as objective chance cannot be fully reduced to mechanics in the statistical mechanical setting, there remains an ineradicable element of logical or epistemic probability in a sense akin to Carnap's built into statistical mechanics. I disagree. The assumptions of stationarity and absolute continuity of the chance or statistical probability measure over phase space in Lebesgue measure are integral to the physical theory and are not part of inductive logic – at least not in the statistical mechanical setting. In my judgement, the physicalisation of probability in statistical mechanics is not equivalent to its annihilation.

Even so, the ergodic theory plays a crucial role in integrating the objective chance hypotheses involved in statistical mechanics with physical theory. To the extent that the condition of metric transitivity is satisfied by statistical mechanical systems, the place-holding status of the assumptions about statistical probabilities or chances is alleviated. Shimony (1975, 391) and von Plato (1985, 56–61) are surely right in suggesting that insofar as we look for an "interpretation" of chance or statistical probability, that is a matter of scientific research – not semantic theory. Consequently, I stand opposed to efforts on the part of the frequentists and assorted modalists and propensity theorists to provide semantic constructions for chance. Insofar as there is any interesting semantics of chance it will derive from the researches in the various disciplines in which probabilistic assumptions are made. It is not to be expected that frameworks for supplying a semantics appropriate to applications in one domain should be appropriate to other domains. What is common to all applications of chance is the specification of a kind of trial, sample space and probability distribution together with principles of direct inference. There is no more a science of probabilities like any other science, as von Mises (1957, 30–31) used to say, then there is a science of dispositions or a science of abilities.

[2] The idea of relativising explanation from statistical premisses to hypothetical knowledge situations obtained via suitable transformations of the actual knowledge situation is proposed by Gärdenfors (1980, 412–413). See also Note 3.

[3] Gärdenfors (1980, p. 405) contrasts Hempel's view according to which the explanans in statistical inductive explanation should show why the explanandum was to be expected with his view that it shows only why it is less surprising. Let K be the current body of knowledge containing the explanandum event E but not the potential explanans T and C. Let $K*$ be the product of conracting K by removing E with minimum loss of information and let $K**$ be the result of adding $T\&C$ to $K*$. E is shown to be less surprising by $T\&C$ if and only if the credal probability of E relative to $K**$ is greater than the credal probability of E relative to $K*$. Gärdenfors' idea does not cohere with any of the explanatory themes I have identified except the case of explanation showing why an event under a description was to have been expected for certain as in deductive nomological explanation. As a general rule, in explanation how possible, the agent's credal probability for E relative to $K*$ will be indeterminate. It may sometimes become

more determinate relative to K^{**} but, as a rule, it will not increase. In general, the state of credal probability judgement relative to K^{*} is irrelevant to assessing what the agent should accept or reject relative to K^{**} and, hence, whether E is shown to be unsurprising or to be expected. See Note 4.

[4] Sometimes probabilists equate maximizing measures of support with (positive monotonic functions of) differences or ratios of posterior and prior probabilities as, for example, Gärdenfors does (see Note 3). My reasons for rejecting such proposals were first given in Levi, 1967b and reiterated in Levi, 1984b. See also, Levi, 1986. Nonetheless, the expected epistemic utility measures I recommend for use as maximizing indices of support may be viewed as generalizations of differences between posteriors and priors. In lieu of priors, I recommend using information determining probabilities and the difference is dampened by an index of boldness q.

[5] Malament and Zabell (1980) are very explicit in relying on the fact that the probability distribution of values of a given phase function are unimodal with small variance and mode equal to the mean of the distribution in recommending that it is to be expected that the value of the phase function will be close to the mean value. They argue that the expectedness is grounded in the high probability that the value of the phase function will be close to the mean value. I do not question that the conclusion they specify is to be expected and that this is related to the fact that the probability distribution is unimodal with small variance and mode equal to the mean of the distribution. But the high probability feature is an unimportant aspect of the problem.

REFERENCES

Coffa, A. (1974) 'Hempel's ambiguity', *Synthese* **28**, 141–164.

Cohen, L. J. (1977) *The Probable and the Provable*, Clarendon Press.

Davidson, D. (1967) 'Causal relations', reprinted in *Actions and Events*, Oxford, pp. 149–162.

Ehrenfest, T. (1958) 'On the use of the notion "Probability" in physics', *American Journal of Physics* **26**, 388–392.

Ehrenfest, P. and Ehrenfest, T. (1959) *The Conceptual Foundations of the Statistical Approach in Mechanics*, translated by M. J. Moravscik, Cornell U. Press.

Gärdenfors, P. (1980), 'A pragmatic approach to explanations', *Philosophy of Science* **47**, 404–423.

Hacking, I. (1965) *Logic of Statistical Inference*, Cambridge U. Press.

Hempel, C. G. (1964) *Aspects of Scientific Explanation*, Free Press.

Kyburg, H. (1961) *Probability and the Logic of Rational Belief*, Wesleyan University Press.

Kyburg, H. (1974) *The Logical Foundations of Statistical Inference*, Reidel.

Levi, I. (1967a) *Gambling with Truth*, reissued in paper by MIT Press (1973).

Levi, I. (1967b) 'Information and inference', reprinted in Levi (1984a), 51–69.

Levi, I. (1976) 'Acceptance revisited', in *Local Induction*, R. Bogdan (ed.), Reidel, pp. 1–71.

Levi, I. (1977a) 'Direct inference', reprinted in Levi (1984a), 192–213.

Levi, I. (1977b) 'Subjunctives, dispositions and chance', reprinted in Levi (1984a), 165–191.

Levi, I. (1979) 'Abduction and demands for information', reprinted in Levi (1984a), 87–106.

Levi, I. (1980) *The Enterprise of Knowledge*, MIT Press.

Levi, I. (1983) 'Truth, fallibility and the growth of knowledge', reprinted in Levi (1984), 109–127.

Levi, I. (1984a) *Decision and Revisions*, Cambridge University Press.

Levi, I. (1984b) note on 'The impossibility of inductive probability', *Nature* **310**, 433–434.

Levi, I. (1986) 'Probabilistic pettifoggery', *Erkenntnis* **25**, 133–140.

Levi, I. and Morgenbesser, S. (1964) 'Belief and disposition', *American Philosophical Quarterly* **1**, 221–232.

Malament, D. and Zabell, S. (1980) 'Why Gibbs phase averages work – The role of ergodic theory', *Philosophy of Science* **47**, 339–349.

Salmon, W. (1984) *Scientific Explanation and the Causal Structure of the World*, Princeton University Press.

Shimony, A. (1975) 'Carnap on Entropy', in *Rudolf Carnap*, *Logical Empiricist* J. Hintikka (ed.) Reidel.

Steiner, M. (1983) 'Under a description', in *How Many Questions?*, Caumen *et al.* (eds.), Hackett, 120–131.

von Mises, R. (1957) *Probability*, *Statistics and Truth*, George Allen and Unwin.

von Plato, J. (1985) *Probability in Dynamical Systems*, unpublished manuscript.

CLARK GLYMOUR

ARTIFICIAL INTELLIGENCE FOR STATISTICAL AND CAUSAL MODELLING

1. PHILOSOPHY AND APPLIED STATISTICS

This is an essay about a computer program, and an "artificial intelligence" program at that. I and several collaborators,[1] who are all philosophers, have developed the program over the last four years. The aim of the program is to assist people in developing statistical theories, or "linear causal models" as they are sometimes called, for non-experimental data. And what has this to do with philosophy?

Philosophical issues that are forced on the physical sciences, in circumstances where they make little or no practical difference, have variants that are of urgent practical importance in the social and behavioral sciences, and in applied statistics generally. For example, philosophers throughout this century have argued that physical theory is underdetermined by all possible evidence; others, including myself, argue to the contrary that when appropriate standards of explanation are imposed, we do not suffer a general inability to decide the best theory on the evidence. The issues are obscure, because they depend on obscure notions about "all possible evidence," and they are of little or no consequence in scientific practice.

In contrast, a related question is perhaps the single most urgent problem in the application of statistics to non experimental data. Consider not all possible evidence, but any fixed body of statistical data. The applied statistician's problem is that there are generally more alternative theories of the data than can possibly be considered; the number of alternatives exceeds the astronomical, and only a few can be analyzed, tested, and compared with each other and with the data. Although the problem seems unphilosophically practical, it is not. Most philosophical issues about the justification of knowledge and belief arise because of some general aspect of our finiteness; our ability to consider alternatives we know to exist is such an aspect.

The problem is not exactly one of underdetermination, but one of discovery: even when there is a best explanation of the data, how can it be discovered among the myriad of possibilities? I think part of the

William L. Harper and Brian Skyrms (eds.), Causation in Decision, Belief Change, and Statistics, II, 223—247.

answer lies in the same ideas about explanation that operate in the physical sciences, and that once we have a clearer idea of what makes for a good causal explanation of statistical data we will know better how to search for such explanations. But given a clear idea of what constitutes a best explanation, the next logical and interesting question is how, given the data and background knowledge, the best theory (or theories) can be located. That question naturally leads to the search for an *algorithm* that will locate the best explanation. And given an algorithm, nothing is more tempting than to install it in a computer and watch it run. Presto, artificial intelligence. I don't quite know where in this the philosophy ends and the computer science begins. I don't see why anyone should care.

2. TOO MANY THEORIES

When we consider statistical models' in the social and behavioral sciences, we find everything from very small models in educational research with as few as six or seven measured variables, to econometric models containing several hundred variables. In its most schematic form, a model for a set of variables postulates a set of causal relationships among the variables. We can if we like think of the causal part of any such theory as given by a directed graph, with the nodes of the graph representing the variables, and directed edges from causes to effects. Each distinct way of arranging a set of variables into a directed graph represents a distinct model of the causal relationships among the variables, and thus an alternative causal theory of the data. The point is that even with a small collection of variables, the number of distinct possible causal arrangements, and thus the number of distinct, alternative causal models of the data, is astronomical. For example, with only six variables, there are 4^{15} different possible directed graphs or causal models.

A social scientist or psychologist or epidemiologist or biologist attempting to develop a good statistical theory has a lot of difficult tasks. He or she must choose what to measure and to measure it, worry about sampling technique and sample size. The researcher must consider whether the variables are multinormally distributed, or have some other distribution. He or she must worry about whether measures of a variable in one individual or place or time are correlated with measures of that same variable in another individual or place or time. The researcher must consider whether the relations among the variables are linear, or non-linear, or even discontinuous. These are demanding tasks, but there are a variety of data analytic techniques to help one along and to test hypotheses concerning these questions. Suppose the researcher has passed these hurdles and has arrived, however tentatively, at several statistical modelling assumptions: the relationships are linear, or close enough to linear, the distribution is multinormal, or nearly so, there is no autocorrelation, or at least not much. Assume that there are covariance data for six variables. The troubles that now face the theorist make the data analysis problems pale in comparison. There are 4^{15} alternative possible theories of the causal dependencies among the six variables, and only one of those theories can be true. How is the one needle in the enormous haystack of possibilities to be found? Perhaps the goal need not be to find the one unique correct theory, but only to reduce the possibilities to a handful. How is even that to be done? The usual thing to say is that the theorist must apply his or her "substantive knowledge" of the domain, but this is mostly whistling in the dark. To get the number of possibilities down to manageable size will take a lot more substantive knowledge than we usually have about social or behavior phenomena. Suppose, for example, the researcher is quite certain that the causal relations do not form cycles. That is, there is no sequence of directed edges in the true graph of the causal relations that leads from one variable back to that same variable. Then the number of alternative causal models consistent with that restriction is still more than three million. Suppose, what is in fact very unusual, that the variables are ordered by time of occurrence, and one therefore knows that variables occuring later cannot be causes of variables occurring earlier. There are still 2^{15} alternative models.

Now repeat the same sequence of calculations when there are twelve variables in the data set. With no restrictions imposed, there are 4^{66} alternative causal models, only one of which can be true. If the researcher knows that there are no cyclic paths in the true model, the number of alternatives is still beyond the astronomical: 521 939 651 343 829 405 020 504 063 (see Harary and Palmer, 1973). If the theorist is lucky enough to be able to totally order the variables and knows that later variables cannot cause earlier variables, the number of alternatives is reduced to 2^{66}.

These counts are conservative. They do not not include the possibility, which every applied statistician considers, that some part of the correlations is due to unmeasured variables that have effects on two or more measured variables. Including such possibilities enormously increases the numbers. What can possibly be done in the face of such an enormous number of possible causal models, no more than one of which can be correct? In practice, even the best practitioners simply take a wild guess, or indulge their prejudices rather than their knowledge. One, or at most two or three, causal models are suggested, appropriate equations are written down, parameters are estimated and the model is subjected so some statistical test. If the test is passed, the researcher is happy and journal editors are happy and the thing is published.

The practice of model building in the social and behavioral sciences looks very much like irrational judgement under uncertainty. The very phenomena that Kahneman and Tversky (Kahneman *et al.*, 1982) describe in experimental subject seem to be exemplified in the practice of social science. Surely the number of alternative models ought to have something to do with the prior probability of any particular model, and in fact the prior probability of any particular linear model in the cases described above ought to be quite small. According to Bayes' theorem, the posterior probability of any particular model should take into account both the likelihood of the evidence on the model, and the prior probability of the model. Current practice is to hit upon a model, somehow, and to advocate the model if it passes appropriate statistical tests. In effect, those who follow this practice are ignoring the prior probabilities, and judging a model to be good enough to accept if it yields a sufficiently high likelihood.

The difficulties are occasionally recognized. In a recent study of data from the summer Head Start program, Jay Magidson (1977) remarks that

> The problem we face is that there is an infinite number of ways to formulate a causal model, and it is not a straightforward matter to determine how to go about doing it, particularly when the causes are unknown and/or unobserved. It is important for researchers to formulate not one but many models so that they can determine whether their conclusions may differ if they accept a different set of assumptions. It is also important to follow some general guidelines in building (formulating) models when the researcher has limited information about the causal process....

An ideal Bayesian solution would be to impose a prior probability distribution on all of the alternative statistical models that are consistent with prior knowledge (or with prior knowledge and some simplifying assumptions), compute the likelihood function for an appropriate statistic, obtain the evidence, and with the likelihood and the prior probability distribution, compute the posterior distribution.[3] The trouble is that no one knows how to do such a thing. Although their have been significant advances in the computation of Bayesian statistics, I think it is fair to say that no feasible algorithm for carrying out a procedure of this sort is known. Perhaps someday such a procedure will be feasible, but what do we do in the meanwhile? Resort to history.

3. FORGOTTEN HISTORIES

The methods that are used in our program, called TETRAD, derive from work in statistics, biometrics and psychometrics that is now long forgotten, but once captured the attention of some of the best workers in these subjects. That work forms a fascinating piece of the history of applied statistics, and reconstructing it may well be the best way to introduce the TETRAD methodology. The story that follows is certainly not complete, especially in its recounting of more recent work.

Charles Spearman was an English psychologist who sought to combine the psychometric innovations of workers such as Binet with the new statistical methods that were emerging from mathematicians such as Pearson. Pearson did not like the result, and he and Spearman remained antagonists for a long while. In 1904 Spearman published, "General Intelligence Objectively Determined and Measured." In it, from a variety of tests with a small sample of London schoolboys. Spearman argued that there is one common component to all manifestations or tests of "intelligence". Reading Spearman's paper nowadays one is struck by the bizzare choice of tests, many of which involved motor skills or sensory discrimination, and the absurd sample sizes, but that is not the point. The point is Spearman's *argument*, for it was the argument that caught the imagination of statisticians and psychometricians.

Spearman's argument, given rather obscurely in this first paper, but more clearly in subsequent papers, was as follows:

Suppose there are four measured variables x_1, x_2, x_3 and x_4, and their correlations (Pearson product-moment correlations, that is) have values that satisfy three equations:

$$\rho_{12}\rho_{34} - \rho_{13}\rho_{24} = 0$$
$$\rho_{12}\rho_{34} - \rho_{14}\rho_{23} = 0$$
$$\rho_{13}\rho_{24} - \rho_{14}\rho_{23} = 0$$

These constraints on the correlations can be explained in a powerful way by supposing that each of the measured variables is affected by a single common unmeasured variable, and that each of the measured variables is also affected by an unmeasured variable specific to it and having no effect on the other measured variables. The algebra is easy:

$$x_1 = a_1G + s_1$$
$$x_2 = a_2G + s_2$$
$$x_3 = a_3G + s_3$$
$$x_4 = a_4G + s_4$$

G, the s_i and the x_i are random variables, the a_i are undetermined constants. Multiplying each side of the first equation by the corresponding side of the second equation and averaging we get:

$$\exp(x_1x_2) = \exp(a_1a_2G^2 + a_1s_2G + a_2s_1G + s_1s_2)$$

For algebraic simplicity assume that all of the variables have mean 0. Since there is assumed to be no causal connection between the several s_i variables, or between these variables and G, their covariances should vanish, leaving

$$\mathrm{Cov}(x_1x_2) = a_1a_2\mathrm{Var}(G)$$

Analogous equations are obtained in the same way for the other covariances of the measured variables. Hence, for example:

$$\mathrm{Cov}(x_1x_2)\mathrm{Cov}(x_3x_4) - \mathrm{Cov}(x_1x_3)\mathrm{Cov}(x_2x_4)$$
$$= a_1a_2a_3a_4\mathrm{Var}(G)^2 - a_1a_3a_2a_4\mathrm{Var}(G)^2 = 0$$

There are three equations of this kind for each foursome of measured variables. The correlation coefficient is just the covariance divided by the product of the standard deviations of the correlated variables, so the difference of the products of the correlations coefficients must also vanish.

Spearman's point was that the assumption of one common factor would explain why these constraints on the correlations, or *tetrad equations* as they came to be called, were satisfied in the population. The thing that must be emphasized, although it is simple enough, is this:

The one factor model implies that the three tetrad equations will hold in the population no matter what the values of the coefficients a_i may be.

The converse is also true, Spearman thought. That is to say, if correlations among four variables satisfy the three associated tetrad equations, then the correlations can be generated from some one factor model, with appropriate values for the linear coefficients. The three tetrad equations are the necessary and sufficient statistical conditions for the admissibility of a one factor explanation of the correlations. This amounts to saying that a one factor model implies no other independent constraints on the population covariance matrix besides the vanishing of all tetrad differences.

Analogous arguments apply for any larger number of measured variables. If there are n measured variables then the number of possible tetrad equations among these variables is 3 times the number of ways of choosing 4 from n. If the correlations are due to the action of one common cause, then all of the tetrad equations must hold, regardless of the values of the linear coefficients and, conversely, if all of the tetrad equations hold, then the correlations can be generated by a one factor model with appropriate values of the linear coefficients.

Spearman realized that the three tetrad equations could be derived from alternative models with multiple latent factors, provided the linear coefficients were chosen to have appropriate values. For example,

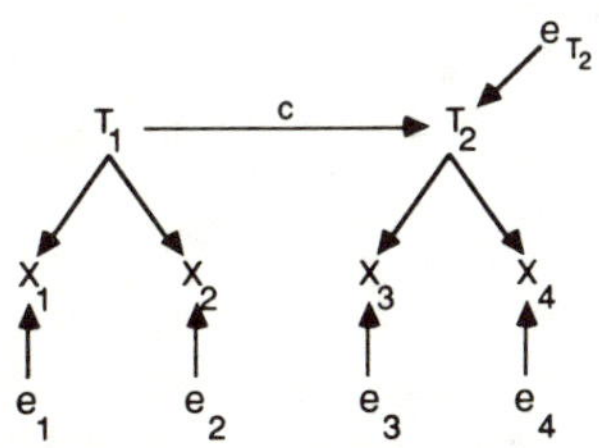

A linear theory whose causal relations are as depicted will only imply one tetrad equation no matter what the value of its linear coefficients, but if the value of the coefficient labeled "c" is unity, then three tetrad

equations must hold. But only the one factor model, he seems to have thought, would imply the tetrad equations no matter what the values of the coefficients might be. Spearman's dominant methodological idea, never fully articulated, seems to have been that the best explanation is one which generates constraints found in the population measures without having to assume special values for its parameters. In other terms, he preferred those explanations which were robust or resilient over changes in the specification of the value of free parameters.

Whether Spearman realized it or not, he was appealing to a methodological idea that runs throughout the sciences. Roughly the same idea has been explicated as resiliency by Brian Skyrms (1980), as simplicity by Roger Rosencrantz (1977) and in still other ways by Robert Riddell (1980), by Robert Causey (1977), and by myself (Glymour, 1980). Kepler gave an analogous argument for the superiority of Copernican to Ptolemaic astronomy (Glymour, 1980). In the same years that Spearman was applying the argument to psychometrics, Harold Jeffreys, who was later to make distinguished contributions to statistics, and Arthur Eddington were making an analogous argument for Einstein's general theory of relativity.

Neither Spearman nor his students nor his critics seem to have noted that there are a number of alternative models that have the same relationship to the tetrad equations as does the one factor model. Pictorially, any of the following models entail that three tetrad equations will hold in the population, no matter what the values of the linear coefficients.

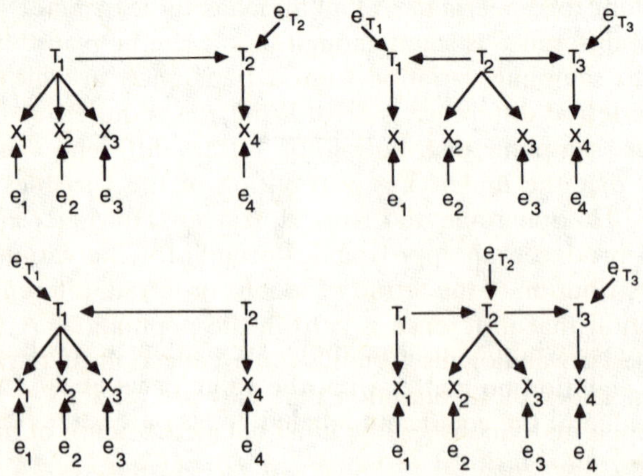

Similar but more numerous alternatives are available when there are more measured variables and all possible tetrad equations hold.

It is hard to know what Spearman would have said about these alternatives. He might have dismissed them on the grounds that they are less simple than the one factor theory, which indeed they are. More likely, he would have held them to be "subtheories" of the one factor model – that is, they amount to saying that the single factor consists of two or three causally connected components. Spearman had nothing against substructure.

In fact, however, the case is even worse, for three tetrad equations among four latent variables are robustly implied by models that have no latent factors, but simply postulate directed effects of the measured variables on one another. For example:

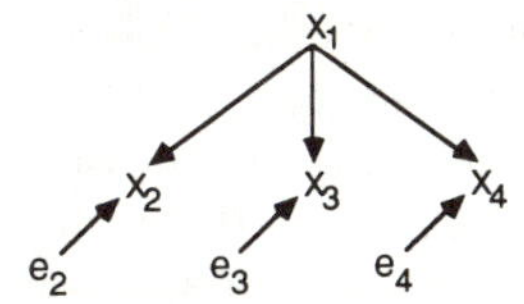

The example is but an instance of Sewall Wright's "path models" (Wright, 1934). Wright's work was published in 1921 and drew immediate attention. So far as I know, however, none of the psychometricians of the day, Spearman included, entertained the thought that the responses their subjects gave to earlier items on batteries of mental tests might have an effect on the responses given to later items.

The increasing appreciation of the importance of sampling error made it urgent for Spearman to find a means to determine whether or not a tetrad difference is small enough that it can be plausibly thought to be due to sampling variation from a population in which the true value of the tetrad difference is zero. What was required was a knowledge of the sampling distribution of tetrad differences under the assumption that the underlying distribution of the variables is multinormal. In 1926 Spearman and Kurt Holzinger (Holzinger and Spearman, 1926) produced an approximate formula for the variance of the sampling distribution of the tetrad residuals, or tetrad difference, under the assumption that difference is zero in the population. A year later they obtained an analogous formula for a collection of tetrad differences, again under the assumption that they are all zero.

In 1927 Spearman published the most detailed defense of his theory, together with the statistical formulae, in *The Abilities of Man* (Spearman, 1927). The book drew a lot of attention, including some unwelcome attention from Spearman's old nemesis, Karl Pearson, Pearson wrote unflattering and anonymous assessments of the book in *Nature*, and then, with Margaret Moul, a critical but substantive contribution in *Biometrika* (Pearson and Moul, 1927). They addressed the question of the sampling distribution for tetrad differences, assuming a vanishing difference in the population and multinormally distributed variables. Pearson did not obtain an exact sampling distribution (and so far as we know the question is still open) and he did not improve on Spearman and Holzinger's formula for the sampling variance. He did, however, observe that the distribution of tetrad differences is certainly not normal. Pearson roundly criticized Spearman, principally on two counts. One, that (using Spearman and Holzinger's formula for the sampling variance) the tetrad residuals in Spearman's data were sometimes large in comparison to the sampling variance. Two, that Spearman had used very small samples (e.g., 50 subjects) and in his statistical analysis had assumed that the sampling distribution of the tetrad differences is normal. Pearson's point was that the sampling distribution is only asymptotically normal, and the sampling variance formula could therefore only be used to decide the significance of a tetrad residual provided the sample sizes were reasonably large, and Spearman's samples weren't. Spearman didn't get the point, for he replied (see Spearman, 1928) that he made no assumption about the normality of the distribution of the tetrad differences. But if so, he could not have made judgements he did about significance.

John Wishart (1928), who had solved the general problem of the sampling distribution of the correlation coefficient, took up the analogous problem for tetrad differences. He did not attempt to find an exact sampling distribution, but he did find an exact formula for the variance of the distribution. The exact formula gives a larger standard deviation than does Spearman and Holzinger's approximate formula, and had the effect of making Spearman's judgements in *The Abilities of Man* look a bit more reasonable.

Spearman was interested in tetrads because he saw in them an argument for his psychological theory. But others saw in his work something much more important, namely a general procedure for finding the best linear causal models to account for statistical data. This ingenious work, pursued in different ways by Maxwell Garnett, Truman Kelley and Karl Holzinger, came to naught, for reasons I will describe subsequently, but it is the real key to our own heuristic search procedures. To explain their ideas, one statistical remark is necessary. Spearman's tetrad equations amount to a simple constraint on the population covariance matrix. For example, to say that all of the tetrad differences vanish is to say that the population covariance matrix has rank one, and that condition proves to be necessary and sufficient for the admissability of a one factor model that provides a robust or resilient explanation of the correlations. Now there are other possible restrictions one could put on the covariance matrix, and in fact the *a priori* possibilities are endless. Garnett and Kelley and in effect posed the following problem:

Find for any linear causal model the constraints on the population covariance matrix necessary and sufficient for the admissibility of the model in the sense that the model will imply the constraints for all values of its linear coefficients, and will not imply any further independent constraints.

This is an elegant problem, and it suggests a general strategy for statistical modelling. If one could classify all of the possible linear models by the constraints necessary and sufficient for them, then by estimating from the sample covariance matrix which constraints are satisfied by the population covariance matrix, one could hope to find the best explanation for any body of data.

Garnett (1919) and especially Kelley, worked out the constraints for a number of models, and the results are reported in Kelley's oddly titled *Crossroads in the Mind of Man* (Kelley, 1928). Kelley's book contains analyses of a number of cases, including, of course, Spearman's one factor model.

Kelley's work was very nice, and in some ways the best done by the psychometricians of the time. As a strategy for locating the best linear causal model, however, it was clearly doomed. There are, just far too many possible cases for a complete classification to be carried out. Each case that generates distinct algebraic constraints requires a distinct sampling statistic, and even the indefatigable Kelley did not obtain sampling variances for some of his cases. Kelley realized as much: "we do not have, and certainly cannot readily obtain, criteria for six, seven or more variables at a time." (*op. cit.*, p. 101).

On the other hand, the cases Kelley did cover offer a kind of *heuristic* guide in the search for models that will best explain the data.

However large the model considered, it can be depicted as a directed graph, and the graph will have subgraphs consisting of only three, or four, or five measured variables, and their connections to each other and to latent (i.e., unmeasured) variables. According to the structure of the subgraphs, constraints of the kinds that Kelley studied will be implied. Thus one might hope to guide the selection of models by comparing the constraints (of the sort Kelley describes) that hold in the data with the constraints of the same sort implied by a hypothetical model, and altering the model if necessary until what holds and what is implied fit together. This is exactly what Kelley himself did.

Kelley applied his analysis to a data set involving nine measured variables for a sample of 140 seventh graders. He settled for trying to fit the model to the vanishing tetrads found approximately in the data. Constraints of higher order were just too troublesome to calculate. Holzinger (1929) took over the same strategy for the same data in defence of a model that was closer to Spearman's views. A year later, in his *Statistical Resume of the Spearman Two Factor Theory* (1930). Holzinger repeated the same strategy on various examples.

Holzinger's *Statistical Resume* was nearly the last gasp of tetrad analyses. Spearman's argument that the best explanation is one that robustly implies the constraints found on the covariance matrix had briefly led to a strategy for scientific discovery. The strategy was to compute the tetrad equations (and other simple constraints) implied by a model, compare them with the constraints satisfied by the data, and alter the model if there was a misfit. That strategy failed, and was carried out in only a few cases. The reasons were computational. The determination of the tetrad equations implied by an arbitrary model, even a model with as few as nine or ten variables, were excessively tedious.

The real death of the analysis of constraints on the covariance matrix came at the hands of Hotelling and Thurstone. In the early 1930s Harold Hotelling developed both principal components analysis and canonical correlation analysis. The two techniques enabled researchers at least to manipulate their data and define constructs which localized the variance of a set of variables, or the correlation of two families of variables. But more important perhaps was Thurstone's contribution.

Thurstone's *The Vectors of Mind* (1931) appeared in the middle of the 1930s, and it effectively ended the analysis of tetrad equations and overidentifying constraints for another quarter century. Despite the denunciations of factor analysis one hears nowadays, Thurstone's book was elegant and revolutionary. Written in a lucid and courteous style, *Vectors of Mind* was for its time a mathematical *tour de force*, completely in command of the relevant matrix theory, geometry and statistics. Equally important, however, it had a simple and appealing philosopical theme – *science is nothing more than simplification* – and it combined that theme with a simple algorithm for doing science as Thurstone conceived it, an algorithm that could be carried out by hand with pencil and paper. The procedure was not tied to any one theory of mental or social phenomena, but to a general conception of science. In Thurstone's view science was nothing more than the erection of "constructs" which describe the data more simply, with fewer "degrees of freedom", and he provided an almost magical procedure by which science could be done, theory discovered, progress made.

Spearman's tetrads were discussed by Thurstone, but as an inferior and inadequate special case. The motivation behind them, and the conception of scientific explanation from which they sprang, was never mentioned. Spearman was a mediocre mathematician, a rude and aggressive controversialist, a man tied to one theory and without an automatic method for discovery; Spearman didn't have a chance. Neither did Kelley and Holzinger, who quickly converted from the analysis of constraints on the covariance matrix to the method of factor analysis.

In almost every way, save that it was an effective and easy procedure for producing theories. Thurstone's factor analysis was inferior to the efforts at constraint analysis pursued by Spearman and Kelley and Holzinger and others. Psychometricians had not entertained the possibility that measured variables might have direct effects on one another, or even on latent variables, nor had they analyzed models in which latent variables had *causal* effects on one another, but nothing in the earlier procedures prohibited such analyses. With factor analysis such relationships were impossible because they were beyond the algorithm, and the philosophical views that went with factor analysis discouraged researchers in psychometrics, and in the many other subjects to which factor analysis quickly spread, from even thinking of the possibility. Thurstone's factor analysis did not, in any serious sense, conduct a search for alternative theories of the data. Instead it churned out a representative of a collection of statistically equivalent models. Worst of all, perhaps, Thurstone's methods abandoned the insight into scientific explanation that had lain behind Spearman's procedures and that had, for a brief while, threatened to bring statistical modelling into the spirit of the natural sciences. For Thurstone, constraints exhibited by the sample covariance matrix had to be satisfied, but they were to be satisfied by adjusting the linear coefficients – the factor loadings – in whatever way necessary. Thurstone was the Ptolemy of statistical modelling.

Thurstone's triumph, and the reasons for it, are evident in Guilford's *Psychometric Methods* (1936). Guilford gives a detailed and sympathetic treatment of Spearman's views and of tetrad methods, before recommending that they be abandoned in favor of Thurstone's factor analysis, which is described in even greater detail. Guilford's reasons for the preference were clear:

> The use of the tetrad is a laborious procedure except when one is dealing with a small number of tests. The number of tetrads increases enormously as the number of tests increases. With 10 tests there are 630 tetrads and 20 tests there are 14535. More effective and economical methods are now available; hence tetrads have little more than historical importance.

Constraints on the covariance matrix came to be known in the economics literature as "overidentifying constraints." For reasons I do not pretend to understand, many econometricians were much more interested in parameter estimation than in theory testing, and overidentifying constraints made consistent, unbiassed estimation more difficult. Some econometricians adopted the exact contrary of the views of the Spearman school, and recommended avoiding models that imposed any constraints on the covariance matrix. Remarks to this effect can be found in Franklin Fisher's distinguished and influential book *The Identification Problem in Econometrics* (1966), and in other econometric works of the same period.

The connection between statistical models and causal notions was reinvigorated by Herbert Simon's work in the 1950s (Simon, 1953), which attempted to derive the causal claims of a statistical model from the identifiability relations of the variables of the model. The analysis had difficulties, but it seems to have given some applied statisticians encouragement in thinking explicitly about the connection between statistical models and the causal conclusions people habitually drew from them. In psychology, Donald Campbell's (Campbell and Stanley, 1963) work on inferring causal order from the correlations of time-lagged variables, and in economics the work of Herman Wold (Wold and Jureen, 1953), had similar effects. The effect was perhaps most pronounced in sociology, a subject which came rather late to statistical methods and causal models. The leading practitioners, thinking the issues through afresh, in effect rediscovered many of the ideas of the Spearman school.

The clearest example is Hubert Blalock's *Causal Inferences in Non-experimental Research* (1961). Blalock reinvented the procedure of examining the constraints on the covariance matrix robustly implied by a linear causal model. The models he considered were not, however, latent factor models of the kind considered by the psychometricians, but were instead path models with no common unmeasured variables, models in fact of the very kind introduced by Sewell Wright. The constraints on the correlation matrix in these cases are not tetrad or pentad equations or the like, but constraints on partial correlations, Blalock makes no reference to any of work done by the Spearman school.

Spearman's ideas were noted by Herbert Costner and O. D. Duncan (in Goddberger and Duncan, 1973), however, who considered the patterns of tetrad equations implied by a number of simple latent variable models in which the latent variables have causal effects on one another and each latent variable affects two or more measured indicator variables. They do ot refer to the work of Kelley or Holzinger, but their basic idea is much the same. Costner, in collaboration with Ronald Schoenberg (Costner and Schoenberg, in Goldberg and Duncan, 1973) even proposed a more or less systematic procedure for revising linear models with latent variables by locating submodels having two causally connected latent variables each with two measured indicators. Such submodels imply a single tetrad equation, which is no longer implied if a measured indicator of one latent variable has a common factor or correlation with a measured indicator of the other latent variable. If the appropriate tetrad equation is not found to hold empirically, Costner and Schoenberg recommend modifying the model accordingly. Costner and Schoenberg's procedure is still occasionally cited, but does not seem to have been much used, probably because it was confined to locating special sorts of erroneous model specifications and was not really algorithmic. Kelley and Holzinger would have liked the work.

The early psychometricians knew how to compute the covariances or correlations implied by an estimated model, given the variances of the exogenous variables and the values of the linear coefficients. With

Kelley's results they knew how to determine the tetrad equations and certain other constraints for special cases. So far as we know, however, they had no general method for computing covariances and correlations when the models were, in contemporary jargon, non-recursive, that is, when the direct graph of the causal model contained a cyclic path leading from one variable back to that same variable. No such models occurred in psychometrics, but they did occur in the economic literature. The general solution to the problem was provided by an outsider, Samuel Mason. His doctoral thesis in 1951 was devoted to the effects of loops in circuits, and it provided a general procedure for computing the correlations or covariances of a recursive or non-recursive model, given the variances of the exogenous variables and the values of the linear coefficients. Mason's work was elaborated for social scientists by David Heise in his *Causal Analysis* (1975).

The moral to this story is that Spearman and Kelley and Holzinger and others had a good idea about scientific explanation and about how to search for good explanations of nonexperimental data. The idea was not entirely lost, and continues on here are there in recent work in statistical modelling. That it never came to fruition was due to computational problems, not to statistical difficulties. What was lacking was a general algorithm that would permit the rapid calculation of the tetrad equations (or other constraints) robustly implied by *any* linear causal model, would compare those constraints with the constraints satisfied by the sample covariance matrix, and would calculate all of the modifications of the model that would help to bring the robustly implied constraints into better agreement with the constraints satisfied by the empirical data. That is a lot, of course, but it is a problem in algorithm design rather than statistical inference.

4. CAUSAL MODELS AND STATISTICAL MODELS

Turning ideas of the Spearman school into an effective procedure for searching for the best statistical models for a data set requires a little philosophical reconstruction. The "statistical models" that abound in the social and behavioral sciences are often obscurely described. The first thing is to get their mathematical structure right.

Linear statistical models consist of linear equations, generally with undetermined coefficients, relating variables that are treated as random. Factor analysis, path analysis, linear regression, and "structural equation models" are of this sort. In applications, such theories are usually accompanied by a body of causal hypotheses which are often only half explicit. Sometimes the causal hypotheses are given explicitly in the form of a directed graph, but more often they are given only by a stylistic convention and by dubbing some variables as "independent" or "exogenous" and others as "dependent" or "endogenous." The stylistic convention is that an affected variable is written on the left hand side of an equation, and the variables that directly affect it are written on the right hand side of the same equation. Usually each equation contains a random "error" term on the right hand side, understood to represent unknown sources of variation in the variable on the left hand side. If one recognizes the convention, a directed graph representing the causal hypotheses can be reconstructed from the equations.

The underlying statistical assumptions that accompany such models are that the linear coefficients are fixed but unknown, and are not random variables. The equations are understood to apply to each member of the population which the theory is supposed to describe. The error terms are assumed to have the same distribution for each individual.[4]

An important statistical assumption of these very common theories is most easily and most generally stated in terms of the directed graph of hypothetic causal relations. Define a *path* in a directed graph to be a sequence of directed edges such that each edge in the sequence, save the first, has as its tail the vertex that is the head of the directed edge preceding it in the sequence. For example, in the graph

the directed edges from A to B and from B to C form a path. Define a *trek* between any two variables, say X and Y, to be either a path from X to Y or a path from Y to X or a pair of paths, one ending in X and the other in Y, whose directed edges have one and only one vertex as a common tail. For example, in the graph

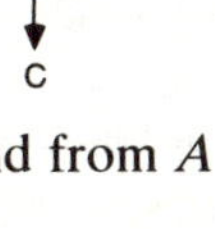

the pair of the from A to B and from A to C form a trek connecting B and C.

The assumption implicit in all theories of these kinds is that if two variables are not connected by a trek, then they are statistically independent.[5]

The directed graph determines everything about a linear causal model save the joint distribution of the variables. It turns out, that a search for alternative causal theories of this sort is essentially a search for alternative directed graphs.

I will leave aside the interesting and controversial questions regarding how the causal claims of such theories are to be understood.

5. THE TETRAD STRATEGY

The search strategy we use and its implementation in the TETRAD program depend on a theorem:

The tetrad equations implied by any linear causal model are determined by the directed graph of the model.

In particular the variances of the variables need not be known, and it does not matter whether the directed graph is cyclic or acyclic. More importantly, we have found an algorithm that computes, for any directed graph and subset of vertices of the graph, the tetrad equations that are implied by any linear causal model whose causal relations are represented by the graph and whose measured variables are represented by the subset of vertices. Still better, the algorithm is fast, and in fact its complexity is low degree polynomial.

The algorithm, implemented in a computer program, enables us to compute almost instantaneously the tetrad equations implied by any linear causal model. The cases so tediously examined by Spearman and Holzinger and Kelley and Duncan and Costner become utterly trivial, and cases which are so complex that no sensible person would try to calculate them by hand are equally easy.

The computer algorithm for computing the tetrad equations implied by any linear model has proved important as a research tool. Imagine trying to carry out the following strategy in searching for good explanations. Start with a simple causal model whose directed graph is a tree. Such graphs typically imply a lot of tetrad equations, only some of which will hold in the data. Determine which of the implied tetrad equations hold well enough in the data and which do not. The addition of further directed edges to a tree will in general *decrease* the number of tetrad equations implied, and in fact the equations implied by the modified graph will generally be a subset of those implied by the original graph. Suppose one knew for each possible addition to the original graph, whether that addition would prevent the implication of tetrad equations that do not hold in the data without preventing the implication of tetrad equations that do hold in the data. With that information one could determine all of the elaborations of the initial theory that would robustly explain all of the empirically correct tetrad equations explained initially, while minimizing the number of incorrect tetrad implications implied.

One inefficient way to get the information required for this strategy is to run our algorithm for computing tetrad equations over and over again for each possible set of additions to an initial graph. That is not a feasible procedure because it consumes resources in way that increases exponentially with the number of vertices in the directed graph. A better strategy is to use the algorithm as a research tool to discover general features of directed graphs that determine how the addition of a direct edge will affect the implication of tetrad equations. That information can then be used as part of a program that guides the search for modifications of an initial model. That investigation was carried out by Richard Scheines, and the results are embodied in the TETRAD program.

TETRAD accepts data and an initial model. It computes the tetrad equations that hold approximately in the data and the tetrad equations implied by the initial theory. The program then searches for all sets of modifications that will preserve the power of the initial theory to explain tetrads that do hold, while reducing the implication of false tetrad equations. The program will find all such modifications, but it is a little too generous and will sometimes include modifications that also prevent the implication of empirically correct tetrad equations. The strategy is essentially that of the Spearman school, but made systematic.

6. RESULTS

Naturally, one wants to know how the program does. It is intended to be run so that the program makes suggestions for modified models and gives the user information to select from among (or add to) those suggestions. The real question is whether people using the program are able to find better theories than people who are not so assisted. Social scientists are reluctant to play John Henry, so we have resorted to weaker tests.

First, we have shown that for most of the linear causal models in the social science literature, especially theories containing unmeasured variables, with the program we can find a number of alternative models of the data which do better by statistical tests or goodness of fit measures.

Second, we have shown that in blind tests with data generated by a model known to the tester but only partially known to a TETRAD user, the latter can reconstruct a great deal of information about the theory from which the data were generated.

Third, we have shown that with good empirical data sets, theories located with the program make dramatic, and correct, predictions.

6.1. BEATING THE SOCIAL SCIENTISTS BY THEIR OWN MEASURES

The point of showing that TETRAD helps amateurs find better theories (by goodness of fit measures) than experts have found without the program is not really to embarass social and behavioral scientists. It is, rather, to show that humans do typically miss a great many alternatives, some of which are relevant as alternative explanations of the data, and that such alternatives can be found with computer assistance.

Example 1: Costner and Schoenberg produced a theory for data about industrialization and political and economic development. Treating the model as the null hypothesis, estimating its coefficients by maximum likelihood, and calculating a chi square statistic yields a statistic with a probability of 0.053, or just barely large enough to escape rejection by the conventional criterion. TETRAD yields two models that add an additional direct edge and have chi square probabilities of 0.141 and 0.294. It also yields a model with two additional directed edges and a chi square probability of 0.276. Somewhat different models are found with chi square probabilities ranging from 0.34 to 0.56.

Example 2: Data from a study by Wheaton and others on the aleination of rural families has been repeated explained by a simple linear causal model. That theory, provided by Joreskog and Sorbom, yields a chi-square statistic with probability 0.335. TETRAD finds two alternative, equally plausible, theories with chi square probabilities of 0.808 and 0.949.

Example 3: The most discussed theory of data from the summer Head Start program fails a standard chi square test. TETRAD finds a theory with chi square probability of 0.30. More importantly, TETRAD finds different models accordingly as it looks at data from Head Start participants or data from Head Start non-participants, and the coefficients of the models differ significantly according to which population they are estimated from. Which strongly suggests that the data should not be explained by structural equation models.

We have obtained similar results using TETRAD on several other data sets from the social science literature.

6.2. BLIND TESTS WITH SIMULATED DATA

Example 1: Data were generated by computer from the following linear theory:

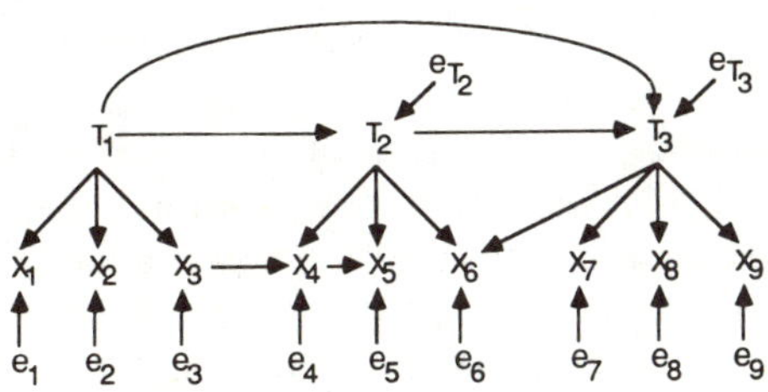

The numbers beside the arrows indicate the values of the linear coefficients. the *T* variables are unmeasured, the *x*s measured, and the *e* variables are "error" terms which were assumed to be distributed normally. A subject, namely me, was given the data (sample size 500) and a description of the initial model, but without the numbers and withut the directed edges $T3 \rightarrow x6$, $x3 \rightarrow x4$, $x4 \rightarrow x5$. I was told that three edges had been deleted from the model, and asked to specify them if I could. Responses could be in terms of sets of alternative models, and in order of decreasing confidence. The number of alternatives is substantial: there are 681 120 different extentions of the partial model, ignoring cases in which arrows go from measured to latent variables and cases in which there is a cycle between two variables. Using TETRAD the alternatives were reduced to a set of 36, which included the correct model. More precise guesses were just that, and were in one way or another incorrect.

Example 2: We generated data with the following model:

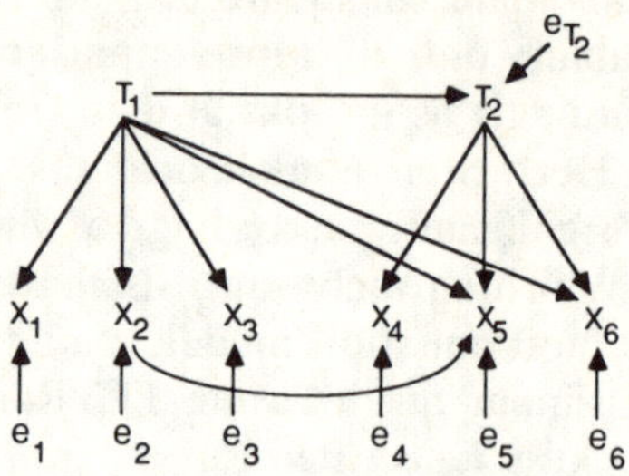

We gave TETRAD the data and an initial model consisting of the one above with the edges $2 \rightarrow 5$, $T1, \rightarrow x5$ and $T1 \rightarrow x6$ deleted. TETRAD will automatically suggest adding *treks* to an initial model. If one considers only treks produced by directed edges between meaured variables or from latent variables to measured variables, there are 1330 possible alternative three trek extentions of the model given to TETRAD. The program produced six alternative extentions, which included the correct one.

Not every case works this well. If a tetrad equation holds in the data because of the particular numerical values of the linear coefficients and not because of the directed graph structure, then TETRAD may make a mistake. In data generated by large models, with lots of linear coefficients, one might expect some tetrad equations to hold for this reason, and so the program will be wrong in some particulars. Sometimes the robust and resilient explanation, the best explanation if you will, is not the correct explanation. This reflection may make one wonder whether the program yields theories for real data that can be confirmed other than by statistical tests. It does indeed.

6.3. PREDICTING THE ORDER OF EVENTS FROM CORRELATIONS

In 1969 Melvin Kohn carried out a survey of the attitudes of over 300 people, who answered a lengthy questionnaire. The correlations among the five questions on the survey were published by Schoenberg and Richtand in 1984. The five questions were all intended to measure "authoritarian" attitudes, and a factor analysis suggested that these questions did indeed have a single common factor. The analysis given by Schoenberg and Richtand suggested that answers to questions 1 and 4 and questions 3 and 5 had some further source of correlation, and the hypothesis embodying that suggestion yields a chi square statistic with probability 0.57.

Two thoughts occurred to us. One is that answers to earlier questions in the survey might have an effect on answers given to later questions. Whether by setting a mood or through a desire for consistency, or some other mechanism. A second thought is that since later questions cannot in this way affect earlier questions, if the hypothesis is correct then we ought to be able to infer something about the order of the questions from the correlations. Since the numbering of the questions in Schoenberg and Richtand's report was not the order of the questions in Kohn's survey, and we did not know the latter order, this test was blind.

There are 120 different orders in which five questions may be asked. Giving TETRAD the simple one factor model and the data, and assuming that any treks to be introduced had to be introduced by adding directed edges between the measured variables, we found a sequence of sets of models to explain the data, each set a proper subset of its predecessor, and each therefore a riskier hypothesis about the order in which the questions were asked. Our most specific hypothesis was that questions 1, 4 and 5 had preceded question 3, which is consistent with 30 possible orderings of the questions. The correct order is 2–1–4–5–3.

CONCLUSION

Perhaps we were lucky, but I think we aren't *just* lucky. Inference to the best explanation works, and the best explanation is often the robust or resilient explanation. Automated philosophy of science works too, and a lot of artificial intelligence is nothing more.

Of course this work doesn't really remove the problems of theoretical excess in the explanation of non experimental data. There remains the problem of alternative initial theorries, about which the TETRAD program does nothing. We have developed a pilot program, GENERATE, that explores the space of possible initial linear models, subject to whatever knowledge about the causal relations among the variables the user may be able to supply. Quite soon one will be able to put in data and prior information, and get out a variety of "best" alternatives, their estimated parameter values, and their statistical properties. Even so, the methods of the program depend on linearity, and the linearity assumptions are quite fragile. There is no reason, so far as I can see, why procedures similar to ours cannot be used for the determination of the appropriateness of the linearity assumption, and for classes of non-linear statistical models.

Carnegia-Mellon University and
University of Pittsburgh

NOTES

[1] Kevin Kelly, Richard Scheines and Peter Spirtes. The program and its philosophical background are described more fully in our forthcoming: *Discovering Causal Structure: Artificial Intelligence for Statistical Modelling.*

[2] The unfortunate practice throughout statistics and much of the sciences is to term any specific theory a "model" and reserve the term "theory" only for the most general and sweeping of hypotheses. I will use "theory" and "model" interchangeably.

[3] This strategy was suggested to us by our colleague, Jay Kadanc.

[4] One can view these stochastic assumptions as resulting from sampling with replacement.

[5] Of course this assumption can be modified, and often is, by explicitly specifying that a pair of variables not connected by a trek in the directed are correlated. But such a specification is statistically equivalent to introducing into the graph a new unmeasured variable with directed edges into the pair of variables said to be correlated. The theory with the modified graphs accords with the text.

REFERENCES

Blalock, Hubert, M. (1961) *Causal Inferences in Nonexperimental Research*, The University of North Carolina Press, Chapel Hill.

Blalock, Hubert, M. (1971) *Causal Models in the Social Sciences*, Aldine Publishing Company, New York.

Campbell, D. and Stanley, J. (1963) *Experimental and Quasi-Experimental Designs for Research*, Rand McNally, Chicago.

Causey, Robert, L. (1977) *Unity of Science*, D. Reidel, Dordrecht, Holland.

Costner, H., Schoenberg, R. (1973) 'Diagnosing indicator ills in multiple indicator models', in Goldberger, A., Duncan, O. (eds.), *Structural Equation Models in the Social Sciences*, Seminar Press, New York.

Fisher, F. (1966) *The Identification Problem in Economics*, McGraw-Hill, New York.

Garnett, J. (1919) 'On certain independent factors in metal measurements', *Proceedings of the Royal Society* **96**.

Glymour, C. (1980) *Theory and Evidence*, Princeton University Press.

Guilford, J. (1936) *Psychometric Methods*, McGraw Hill, New York.

Harary, F. and Palmer, E. (1973) *Graphical Enumeration*, Academic Press, New York.

Heise, D. (1975) *Causal Analysis*, John Wiley & Sons, New York.

Holzinger, K. and Spearman, C. (1926) 'Note on the sampling error of tetrad differences', *British Journal of Psychology* **16**, 86–99.

Holzinger, K. (1929) 'On tetrad differences with overlapping variables', *Journal of Educational Psychology* **20**, 91–97.

Holzinger, K. (1930) *Statistical Resume of the Spearman Two-Factor Theory*, University of Chicago Press, Chicago.

Kahneman, D., Slovic, P., and Tversky, A. (eds.) (1982) *Judgement Under Uncertainty: Heuristics and Biases*, Cambridge University Press.

Kelley, T. (1928) *Crossroads in the Mind of Man*, Stanford University Press, Stanford.

Kohn, M. (1969) *Class and Conformity*, Dorsey Press, Homewood. Il.

Magidson, J. (1977) 'Toward a causal model approach for adjusting for preexisting differences in the nonequivalent control group situation', *Evaluation Quarterly* **1**, 399–420.

Pearson, K. and Moul, M. (1927) 'The mathematics of intelligence, I. The sampling errors in the theory of a generalized factor', *Biometrika* **19**, 246–291.

Riddell, R. C. (1980) 'Parameter disposition in pre-Newtonian planetary theories', *Archive for History of Exact Sciences* **23**, 87–157.

Rosencrantz, R. (1977) *Inference, Method and Decision*, D. Reidel, Dordrecht, Holland.

Schoenberg, R. and Richtand, C. (1984) 'Application of the EM method', *Sociological Methods and Research* **13**, 127–150.

Simon, H. (1953) 'Causal ordering and identifiability', in *Models of Discovery*, pp. 53–80. D. Reidel, Dordrecht, Holland.

Skyrms, Brian (1980) *Causal Necessity*, Yale University, New Haven.

Spearman, Charles (1904) 'General intelligence objectively determined and measured', *American Journal of Psychology* **15**, 201–293.

Spearman, C. (1927) *The Abilities of Man*, Macmillan, New York.

Spearman, C. (1928) 'Pearson's contribution to the theory of two factors', *British Journal of Psychology* **19**, 95–101.

Thurstone, L. (1931) 'Multiple factor analysis', *Psychological Review* **38**, 406–427.

Thurstone, L. *The Vectors of Mind*.

Wishart, J. (1928) 'Sampling errors in the theory of two factors', *British Journal of Psychology* **19**, 180–187.

Wold, H. and Jureen, L. (1953) *Demand Analysis*, Wiley, New York.

Wright, S. (1934) 'The method of path coefficients', *Annals of Mathematical Statistics* **5**, 161–215.

INDEX

THE UNIVERSITY OF WESTERN ONTARIO SERIES IN PHILOSOPHY OF SCIENCE

A Series of Books in Philosophy of Science, Methodology, Epistemology, Logic, History of Science, and Related Fields

1. J. Leach, R. Butts, and G. Pearche (eds.), *Science, Decision and Value.* 1973, vii + 219 pp.
2. C. A. Hooker (ed.), *Contemporary Research in the Foundations and Philosophy of Quantum Theory.* 1973, xx + 385 pp.
3. J. Bub, *The Interpretation of Quantum Mechanics.* 1974, ix + 155 pp.
4. D. Hockney, W. Harper, and B. Freed (eds.), *Contemporary Research in Philosophical Logic and Linguistic Semantics.* 1975, vii + 332 pp.
5. C. A. Hooker (ed.), *The Logico-Algebraic Approach to Qunatum Mechanics.* 19775, xv + 607 pp.
6. W. L. Harper and C. A. Hooker (eds.), *Foundations of Probability Theory, Statistical Inference, and Statistical Theories of Science.* 3 Volumes. Vol. I: *Foundations and Philosophy of Epistemic Applications of Probability Theory.* 1976, xi + 308 pp. Vol. II: *Foundations and Philosophy of Statistical Inference.* 1976, xi + 455 pp. Vol. III: *Foundations and Philosophy of Statistical Theories in the Physical Sciences.* 1976, xii + 241 pp.
7. C. A. Hooker (ed.), *Physical Theory as Logico-Operational Structure.* 1979, xvii + 334 pp.
8. J. M. Nicholas (ed.), *Images, Perception, and Knowledge.* 1977, ix + 309 pp.
9. R. E. Butts and J. Hintikka (eds.), *Logic, Foundations of Mathematics, and Computability Theory.* 1977, x + 406 pp.
10. R. E. Butts and J. Hintikka (eds.), *Foundational Problems in the Sspecial Sciences.* 1977, x + 427 pp.
11. R. E. Butts and J. Hintikka (eds.), *Basic Problems in Methodology and Linguistics.* 1977, x + 321 pp.
12. R. E. Butts and J. Hintikka (eds.), *Historycal and Philosophical Dimensions of Logic, Methodology and Philosophy of Science.* 1977, x + 336 pp.
13. C. A. Hooker (ed.), *Foundations and Applications of Decision Theory.* 2 volumes. Vol. I: *Theoretical Foundations.* 1978, xxiii + 442 pp. Vol. II: *Epistemic and Social Applications.* 1978, xxiii + 206 pp.
14. R. E. Butts and J. C. Pitt (eds.), *New Perspectives on Galileo.* 1978, xvi + 262 pp.
15. W. L. Harper, R. Stalmaker, and G. Pearce (eds.), *Ifs. Conditionals, Belief, Decision, Chance, and Time.* 1980, ix + 345 pp.
16. J. C. Pitt (ed.), *Philosophy in Economics.* 1981, vii + 210 pp.
17. Michael Ruse, *Is Science Sexist?* 1981, xix + 299 pp.
18. Nicholas Rescher, *Leibniz's Metaphysics of Nature.* 1981, xiv + 126 pp.

19. Larry Laudan, *Science and Hypothesis.* 1981, x + 258 pp.
20. William R. Shea, *Nature Mathematized.* Vol. I., 1983, xiii + 325 pp.
21. Michael Ruse, *Nature Animated.* Vo. II, 1983, xiii + 274 pp.
22. William R. Shea 9ed.), *Otto Hahn and the Rise of Nuclear Physics.* 1983, x252 + 23. pp.
23. H. F. Cohen, *Quantifying Music.* 1984, xvii + 308 pp.
24. Robert E. Butts, *Kant and the Double Government Methodology.* 1984, xvi + 339 pp.
25. James Robert Brown (ed.), *Scientific Rationality: The Sociological Turn.* 1984, xiii + 330 pp.
26. Fred Wilson, *Explanation, Causation and Deduction.* 1985, xviii + 385 pp.
27. Joseph C. Pitt (ed.), *Change and Progress in Modern Science.* 1985, viii + 398 pp.
28. Henry B. Hollinger and Michael John Zenzen, *The Nature of Irreversibility.* 1985, xi + 340 pp.
29. Kathleen Okruhlik and James Robert Brown (eds.), *The Natural Philosophy of Leibniz.* 1985, viii + 342 pp.
30. Graham Oddie, *Likeness to Truth.* 1986, xv + 218 pp.
31. Fred Wilson, *Laws and Other Worlds.* 1986, xv + 328 pp.
32. John Earman, *A Primer on Determinism.* 1986, xiv + 273 pp.
33. Robert E. Butts 9ed.), *Kant's Philosophy of Physical Science.* 1986, xii + 363 pp.
34. Ian B. MacNeill and Gary J. Umphrey (eds.), Vol. I, *Applied Probability, Stochastic Processes, and Sampling Theory.* 1987, xxv + 329 pp.
35. Ian B. MacNeill and Gary J. Umphrey (eds.), Vol. II, *Foundations of Statistical Interence.* 19vii, x + 287 pp.
36. Ian B. MacNeill and Gary J. Umphrey (eds.), Vol. III, *Time Series and Econometric Modelling.* 1987, xix + 394 pp.
37. Ian B. MacNeill and Gary J. Umphrey (eds.), Vol. IV, *Stochastic Hydrology.* 1987, xv + 225 pp.
38. Ian B. MacNeill and Gary J. Umphrey (eds.), Vol. V, *Biostatistics.* 1987, xvi+283 pp.
39. Ian B. MacNeill and Gary J. Umphrey (eds.), Vol. VI, *Actuarial Science.* 1987, xvi + 250 pp.
40. Nicholas Rescher, *Scientific Realism.* 1987, xiii + 169 pp.
41. Brian Skyrms and William L. Harper (eds.), Vol. I, *Causation, Chance, and Credence.* 1988, xii + 284 pp.
42. William L. Harper and Brian Skyrms (eds.), Vol. II, *Causation in Decision, Belief Change and Statistics.* 1988, xix + 252 pp.